Sustainable Action in Economy and Society

Malte Faber · Reiner Manstetten ·
Marco Rudolf · Marc Frick · Mi-Yong Becker

Sustainable Action in Economy and Society

Orientation for Change

Springer

Malte Faber
Universität Heidelberg
Heidelberg, Germany

Reiner Manstetten
Universität Heidelberg
Heidelberg, Germany

Marco Rudolf
Hochschule Pforzheim
Pforzheim, Germany

Marc Frick
Universität Basel
Basel, Switzerland

Mi-Yong Becker
Hochschule Bochum
Bochum, Germany

ISBN 978-3-662-69121-2 ISBN 978-3-662-69122-9 (eBook)
https://doi.org/10.1007/978-3-662-69122-9

Translation from the German language edition: "Nachhaltiges Handeln in Wirtschaft und Gesellschaft" by Malte Faber et al., © The Editor(s) (if applicable) and The Author(s), under exclusive license to Springer-Verlag GmbH, DE, part of Springer Nature 2023. Published by Springer Berlin Heidelberg. All Rights Reserved.

This book is a translation of the original German edition "Nachhaltiges Handeln in Wirtschaft und Gesellschaft" by Malte Faber et al., published by Springer-Verlag GmbH, DE in 2023. The translation was done with the help of an artificial intelligence machine translation tool. A subsequent human revision was done primarily in terms of content, so that the book will read stylistically differently from a conventional translation. Springer Nature works continuously to further the development of tools for the production of books and on the related technologies to support the authors.

This Springer imprint is published by the registered company Springer-Verlag GmbH, DE, part of Springer Nature.
The registered company address is: Heidelberger Platz 3, 14197 Berlin, Germany

Foreword to the Readers

The book you are currently reading is a map for navigating through difficult terrain. It provides an overview and offers landmarks for those striving towards a sustainable future.

In the discourse on environmental and climate protection, there is no shortage of committed participants. Actors from politics, business, the innovation and start-up scene, science, and civil society contribute their perspectives, while activists ensure that the urgency of the topic remains present. The emphatic nature by the protest of the "Last Generation" indicates what it is about: Much needs to change. And as soon as possible. It is not only minorities from the environmental movement who are striving for creative ideas and new solution approaches. Large organizations like the German Energy Agency (dena), of which I was the chairman of the management board for over eight years, repeatedly bring models, scenarios, projects, and suggestions for action options into the debate, and parliamentary majorities pass laws that effectively promote sustainable action. Exceptionally ambitious climate targets have even become legally binding under international law through the Paris Climate Agreement ratified by almost all governments. And yet, there is often a peculiar sense of disorientation.

One reason for this is that we often want too much and perceive too little. Ideas, initiatives, and actions fall flat if they cling only to a single perspective. Too many of us believe we know all too well exactly how everything can succeed. We just need to do this or that! However, the assured conviction of one is often not compatible with the beliefs of others. Especially those who are particularly committed—I do not want to exclude myself here—are exposed to the danger of a certain hubris. But those who only see the natural factors, underestimate the difficulty of societal developments. Those who only pay attention to the capability of initiatives and ideas to achieve a majority, easily forget that nature does not wait, and it backfires if one does not consider the inherent laws of processes in economy and politics. In today's sustainability discourse, many participants are somehow correct, and yet they talk past each other or even fight each other because what is 'correct' for one side is not perceived or understood by the other side. Effective environmental and climate policy is more than just translating scientifically derived goals into a seemingly straightforward political concept. It doesn't help: Those who want to be successful need to address diverse perspectives and questions. Without an overview of the often complex relationships between natural conditions, economic processes, societal trends, and political process structures, we cannot meet the tasks ahead of us.

In order to lay out the field of environmental and climate policy issues in all its breadth and complexity for the readers, this book presents concepts from physics, biology, economics, and political science within the context of ethical and philosophy of science considerations. On the way to a sustainable society, these concepts can structure the discussion and provide orientation. I would like to briefly highlight some of these concepts here. In this book, a prominent theme is time. Transformation processes take time—for the construction of infrastructure, the change of behaviors, the establishment of institutions, the scaling of innovations for example. But also important for the understanding of time is a sense for time windows within which developments can be accelerated. A citizens' movement, for example, can open such a time window, as can the tragedy of a war, which brings awareness of entirely new possibilities for action along with its unprecedented challenges. Another topic is justice. We like to talk about it when it comes to climate protection, but often find ourselves lost in debates about technical, economic, and legal instruments. The book makes it clear that achieving sustainability will not be possible without a common understanding of a fair and just framework. Just as often, we talk about the conversion of energy and the need for resources without being sufficiently aware of the physical framework conditions of their use. However, the treatment of topics such as entropy and the laws of thermodynamics in this book can give us clues to the respective states of energy that we need to understand for a focused approach to these questions. I was particularly impressed by the book's discussions on the topic of ignorance. The knowledge of our own ignorance is one of the most important foundations for the success of our efforts towards sustainability. Too often, we assume we already know exactly what steps will be necessary in 10 or 15 years.

During my time as Chairman of the Management Board of the German Energy Agency (dena), a key principle was that while energy transition and climate protection require good planning and organization, our ignorance must be incorporated into all planning. One must always be open to changes, new options, and solutions. Whether in the dena pilot studies, in which we sought the best ways for the integrated energy transition and climate neutrality across all sectors, or in the dena network study, in which we outlined and developed the foundations for integrated infrastructure planning: All our approaches have always tried to base solution proposals on a secure foundation for actions while allowing flexibility on this foundation to be open for even better solutions in the future, which we could neither foresee nor expect at the outset of our deliberations. To me, robustness and flexibility are not cotradictory, but the basis for success.

Again and again, in the conception of our activities and studies with my team, I have come across the foundational work of the so-called *Heidelberg School of Ecological Economics*, upon which the book before you is based. The founder of this school and one of the authors of this book is Malte Faber. More than forty years ago, he encountered obvious deficits in the pure doctrine of economics. Throughout his life, he has gradually expanded the horizon, developed concepts and ideas in exchange with natural scientists and philosophers beyond the framework of economics, and applied them in dealing with actual problems. He has also — and very successfully — ventured into practical political consulting to test his own approaches. For 30 years, he held the Chair of Economic Theory at Heidelberg University and spent a good part of this time as director of the Institute for Interdisciplinary Environmental Research, which he founded. Many of those who are now intensively involved with the energy transition and climate protection have gone through his school. I am one of them. As someone who has repeatedly benefited from conversations with Malte Faber and the other authors of this book and their references to texts and groundbreaking thoughts, I am particularly pleased about this book. I am convinced that this new, condensed, and practical view of the topics of sustainability, environmental protection, and climate protection comes at exactly the right time.

Berlin
July 2024

Andreas Kuhlmann
Managing Director and expert for
climate protection and energy transition at
Christ & Company. Chairman of
the Management Board of
the German Energy Agency dena from 2015 to 2023

Preface and Acknowledgements

This book originated from Malte Faber, who held the Chair of Economic Theory at the Ruprecht-Karls-University Heidelberg from 1974–2004. It is based on the findings of an interdisciplinary chair team that, in varying compositions and over 40 years, has researched the interactions and dynamics between human actions, the economy, and nature. The findings have been published in numerous publications in journals and publishers of very different disciplines, partly as results of dedicated research projects. In retrospect, however, the findings form an independent research approach. In order to make this recognizable as such and adaptable beyond the Heidelberg group, the approach was translated into a *digital knowledge map* as part of the project *MINE—Mapping the Interplay between Nature and Economy* and thus was made accessible to a wide readership. The map, published in 2019 as the online tool *MINE* (www.nature-economy.com), brings together 15 fundamental concepts, assigns each concept a place within a network with numerous cross-connections, and presents them in an understandable way. The in-depth experiences and teaching successes that the author team was able to achieve in numerous teaching events with MINE provided the impetus to prepare *MINE* didactically and to provide it in book form. Particularly groundbreaking were the numerous valuable hints from participants in the MINE seminars at several universities in Europe, India, and Argentina.

The individual chapters of this book discuss step by step the concepts of *MINE* and refer to the work of researchers who have shaped this specific approach. Initially, it is acknowledged that numerous other scientists in other places and with their own approaches have worked on the same questions and concepts. However, providing an overview of the current state of research on the multitude of questions addressed would exceed the scope of this book and, in our view, also make access more difficult. Therefore, we have chosen to only include the necessary literature references in the chapters themselves and at the end of Parts 1, 2, 3 and 4 of the book, we provide some literature recommendations that go beyond our own work.

This book is the result of the collaboration of an author team from five different generations. Even though we pursue a common goal and deal with the same questions, each member of our group brings their own perspectives and their own approaches to solutions. We have never understood the discussions that arise from the heterogeneity of our group in terms of our age, our life realities, and experiences as an obstacle, but have always experienced the friction and even the dissent that sometimes arose from it as something constructive. We have tried to not hide these different perspectives and opinions in our joint seminars and to keep the struggle for the best ideas and arguments among ourselves, but also in exchange with the students transparent and open-ended. Even though this approach requires more effort than quickly reaching a minimal consensus and even though it comes with impositions and necessitates a basic trust in the respective counterpart, we have experienced it as valuable. The questions and challenges that concern us are important and it is worth constructively arguing about the best ways to solutions. We hope that we have succeeded in conveying this attitude in the present book.

Important to Know: Structure of the Book

- **Note boxes:** Throughout the book, note boxes are incorporated that highlight key aspects of the respective chapters. At the beginning of each chapter, under the title "What's it about?", a concise summary of what the chapter covers is provided. Within the chapters, "Important to know boxes" highlight crucial steps of knowledge and definitions. Some chapters also include a 'Summary' section, where the most important findings are briefly summarized again.
- **Bibliographies:** Literature references are compiled at the end of each chapter in a bibliography.
- **Literature recommendations:** Key literature recommendations for further reading are provided for parts 1–4 of the book. These refer to literature by other authors who address the questions of our book using their respective approaches.
- **Gender rules:** For improved readability, the book alternates between the generic masculine and the generic feminine, with all examples fundamentally referring to all genders, including non-binary ones.

The development of a book is, even with a five-member author team, a collective work that requires a multitude of helping hands and thinking heads beyond the individual group member. We would like to thank the participants of our seminars at the Ruprecht-Karls-University Heidelberg, the Bochum University of Applied Sciences, and the Heidelberg School of Education for the stimulated discussions of the contents of this book and their open feedback on the didactic concept. We would like to express our heartfelt thanks to Springer Verlag and Renate Scheddin for their willingness to include the book in their program and for their helpful and supportive supervision from the manuscript to the print. Many thanks to the colleagues at the Ruprecht-Karls-University Heidelberg, the ZEW-Leibniz Centre for European Economic Research in Mannheim, the Pforzheim University of Applied Sciences, and the Bochum University of Applied Sciences with whom we have discussed the approach and concepts of this book in recent years. Ute Beckel-Faber, Monika Kloth-Manstetten, Jörg Hüfner, Klaus Jacobi, Andreas Kuhlmann, Philipp Krohn, and Thomas Petersen have accompanied not only this book, but practically all projects of the past years as conversation partners, critical readers, and impulse givers. They deserve our special thanks.

Basel, Bochum and Heidelberg
in July 2024

Malte Faber
Reiner Manstetten
Marco Rudolf
Marc Frick
Mi-Yong Becker

Contents

About the Authors

Mi-Yong Becker, Born in 1970, Professor of Sustainability with an economics focus at Bochum University of Applied Sciences. In 2004, she received her doctorate in economics from the University of Heidelberg. From 2004 to 2007, she was a post-doctoral fellow of the DFG at the graduate school "Global Challenges" at the University of Tübingen. Subsequently, until 2019, she was a research associate at the Helmholtz Centre for Environmental Research, most recently as head of the "Governance and Innovation" working group. Since 2018, she has been a recipient of the German Environmental Award of the DBU, together with M. Hirschfeld, R. Müller and M. v. Afferden, for the development and political implementation of a decentralized wastewater system solution in Jordan. Since 2019, she has been a professor at Bochum University of Applied Sciences and since 2022, a member of the presidency as Vice President for Sustainability, Transfer Entrepreneurship.

Malte Faber, Born in 1938, he is a professor emeritus of economics at the University of Heidelberg. He received his doctorate in 1969 from the Technical University of Berlin and qualified as a professor there in 1973. From 1974 to 2004, he held a chair in economic theory at the University of Heidelberg, where he was the director of the Alfred-Weber-Institute and from 1997 to 2004, the director of the Interdisciplinary Institute for Environmental Economics. Since 2004, he has been a professor emeritus and conducts seminars on environmental issues and economic philosophy at German and international universities. From 1981 to 2017, he was an advisor to the German government, the US environmental agency, and the People's Republic of China. Since 2003, he has been teaching courses and giving lectures on "Zen meditation and Christian contemplation" at the University of Heidelberg and other locations.

Marc Frick, Born in 1990, he is an Academic Associate and project coordinator in the General Secretariat of the University of Basel. In 2020, he finished his doctorate in political theory with a book on the social-theoretical implications of gift-giving practices. He published on ecological economics in cooperation with Malte Faber and Reiner Manstetten, as well as on social and ethical aspects of environmental and climate policy. Since 2015, he has been the voluntary chairman of an amateur sports club, an experience that significantly shapes his view on social cohesion and the cooperation of very different people in solving societal challenges.

Reiner Manstetten, Born in 1953, lecturer in philosophy at the University of Heidelberg, teacher of Christian contemplation. In 1992, he received his doctorate in philosophy on Meister Eckhart, in 1997 he habilitated in economics on the concept of man in economics, in 2003 he received the Ernst Bloch Award from the city of Ludwigshafen. Publications on philosophical mysticism as well as on ecological economics and the fundamentals of economics, often in cooperation with Malte Faber. In 2018, the monograph: The Dark Side of the Economy (Alber Verlag) was published.

Marco Rudolf, Born in 1994, academic staff member at Pforzheim University, doctoral student at KIT Karlsruhe. After studying mathematics in Heidelberg and a master's degree in Resource Governance in Chile, he has been researching the topic of hydrogen in the energy transition and corporate emission accounting since 2020. For him, transformation to a sustainable society means not only technical change, but also social changes. In addition to his scientific work, he is committed to various youth organizations and organizes holiday camps for adults, where the focus is on communal group experiences in an environment shaped by sustainable consumption and closeness to nature.

Part I
Starting Points

1 Introduction

Books on climate, the environment, and resource issues are currently popular. Countless articles illuminate the overarching theme of "sustainability" from various perspectives, passionately debate key questions, and confidently present diverse approaches to solving what may be the "greatest societal challenge of our time". This raises the question of whether anything new can still be said on these topics. As a team of authors, we also questioned the necessity of writing this book. Much of its content has already been published elsewhere by others or even by us, the authors. However, we hope that the overall result is indeed novel, as it brings together perspectives and insights under the comprehensive aspect of sustainability. These perspectives and insights essentially belong together but are usually only accessible separately and scattered across various publications in different locations.

What distinguishes this book from others, namely the diversity and sequence of the problems presented, may seem unusual to some. The range of topics extends from natural sciences to economics to philosophy: From physics, we learn about thermodynamics and entropy; from biology, evolution; the topics of joint production and scarcity belong to the field of economics; practical philosophy is represented by investigations into problems such as responsibility, power of judgement, and justice. We believe that these seemingly disparate topics are interconnected, each addressing an aspect that needs to be considered in the forthcoming transformations. Some of these aspects are even at the heart of the expected and hoped-for major changes towards a *sustainable development*. The intended audience of our considerations consists of individuals who wish to participate in societal and political transformation processes - whether in the political arena, administrative bodies, research and development departments of companies, scientifically based advisory processes of expert committees, or various forms of citizen participation and protest. These individuals are addressed, provided they wish to gain

M. Faber et al., *Sustainable Action in Economy and Society*,
https://doi.org/10.1007/978-3-662-69122-9_1

an overview of the range and complexity of the upcoming tasks and are seeking promising intervention possibilities.

We begin the book with a political-philosophical examination of the possibilities for changing our world. In doing so, we integrate the perspectives of economics, ecology, and philosophy (Part 1). This is followed by an exploration of humanity and the concepts with which human action and its effect in the world can be reflected and weighed between alternative courses for action (Part 2). The third part focuses on elementary concepts from the natural sciences that form the basis for understanding the challenges that current environmental problems pose to humanity. The fourth part brings together the different elements of this book by examining essential aspects of the interplay between humanity and nature, and particularly reflecting on the problem of ignorance. In Part 5, when we examine the unity and incompatibility of coexistence between humanity and nature, we indeed encounter limits to human planning and action. However, it also becomes evident that significant transformations are indeed possible.

The MINE Project and the Heidelberg School of Ecological Economics

2

Contents

In this book, we aim to integrate concepts from various disciplines into a unified picture. This approach may be unfamiliar to many readers. However, we believe this comprehensive overview is crucial for developing an appropriate understanding of environmental problems, which is a prerequisite for finding suitable solutions in practice. We have repeatedly observed a lack of basic interdisciplinary approaches. While individual scientists and experts may be highly qualified to investigate specific aspects, there is often a reluctance and lack of time to confront the challenges presented by the bigger picture. This reluctance is understandable, given the apprehension that the challenges of this bigger picture could prove to be an overwhelming task. However, one must confront some of this overwhelming task when considering perspectives that should lead to action. A problem like climate change is not solely physical, biological, economic, or social—it encompasses all these aspects and even more if one truly wants to understand it. Anyone who wishes to contribute to meaningful solutions must integrate all these aspects in some way. However, the overview in this book does not claim to offer the complete big picture. This would be impossible due to the inherent ignorance that no one can evade. Instead, we aim to highlight the complexity of the problems on the one hand and provide orientation on how to consider relevant aspects within this complexity, despite the inevitable ignorance that accompanies all acquisition of knowledge, and prepare for action on the other hand.

M. Faber et al., *Sustainable Action in Economy and Society*,
https://doi.org/10.1007/978-3-662-69122-9_2

The first attempt at such an overview has been available since 2019, not in the form of a book, but rather in the somewhat unusual form of a digital knowledge map. As part of the MINE project, we have attempted to develop a digital map to illustrate the interplay between nature and economy. MINE stands for "Mapping the Interplay between Nature and Economy". The MINE project, like this book, strives to provide an overview of many relevant aspects of environmental problems and to offer guidance. Given the aim to provide orientation, it is not surprising that we have chosen the form of a map in the MINE project (Fig. 2.1).

What is the connection between the MINE project and this book? Both are based on two aspects that serve as the foundation of our thinking:

1. The economy is understood as a subsystem of nature.
2. Questions are more important than answers.

We will explain both aspects in the following sections.

2.1 The Economy as a Subsystem of Nature

Important to Know

Mainstream economics views nature as a subsystem of the economy, providing resources and absorbing emissions, wastewater, and waste. In contrast, this book, following the tradition of ecological economics, perceives the economy as a subsystem of nature.

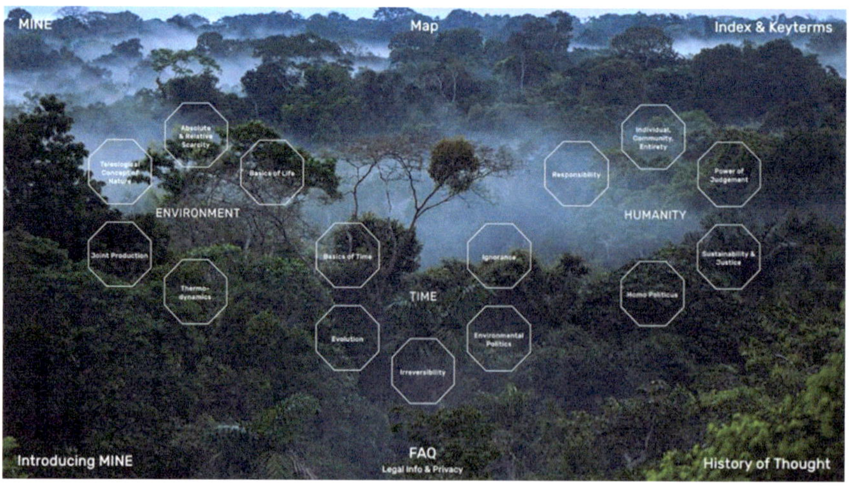

Fig. 2.1 Concepts for investigating the interaction between humans and the environment

Many environmental economic approaches present nature as an element in the regulatory system of the economy, discussing only its use, exploitation, and at most, its resistance. In contrast, we propose that the economy is fundamentally dependent on nature, subject to its laws and embedded within it. In this book, we are primarily interested in the dynamics of nature, which is why we systematically consider nature in connection with time from the outset (Fig. 2.2).

This shift in perspective compared to mainstream economics situates our book within a scientific school of thought known as ecological economics. The understanding of the economy as a subsystem of nature is characteristic of this school of thought, which can be traced back to the economist Kenneth Boulding (1910–1993) and his essay "The Economics of the Coming Spaceship Earth" (1966).

"The *Spaceship Earth* was a common metaphor in the sixties for a world reaching its limits. For Boulding, it also meant that one person's environmental problems can quickly become environmental problems for their neighbour. His text is a founding document of ecological economics, a scientific discipline that, simply put, does not perceive nature as a subsystem of the economy like environmental and resource economics, but the economy as part of the global ecosystem. The best-known representatives are Herman Daly, Robert Costanza, Joan Martinez Alier, and the Heidelberg School around Malte Faber" (Krohn, 2021, para. 6).

The starting point of the MINE project is at the Chair of Economic Theory at the University of Heidelberg. Around Malte Faber, who held this chair from 1974 to 2004, a group of scientists with regularly changing composition has been researching since the early 1980s. They aimed not only to develop theoretical economic insights but also to apply them in practice. Their goal was to make a difference in the face of pressing environmental problems. It quickly became clear that the insights of economics alone would not be sufficient for this purpose. To bridge the gap from theory to practice and to solve complex environmental problems, a research approach was developed that facilitated exchange among people from

Fig. 2.2 **a)** Traditional economics' understanding: Environment as a subsystem of the economy. **b)** Ecological economics' understanding: Economy as a subsystem of the environment.

very different disciplines and created spaces for collaborative and trustful work, drawing on the strengths of different perspectives.

The exchange between science and practice repeatedly led to constructive debates about specific questions and environmental policy in general, which also extended into the political sphere. The engagement with different scientific concepts made it clear that the approaches of conventional economics were not sufficient to understand environmental and resource problems. Instead, energy issues became more important, leading to an engagement with thermodynamics. This combination of economic and physical questions opened the door to further interdisciplinary collaborations. It also became clear that concepts from philosophy, and especially philosophical ways of questioning, are significant for understanding societal transformations.

In this book, essential thoughts are presented in a highly condensed form, as they have been developed over more than 40 years of work within the described research approach. Readers who wish to delve deeper into the contents of the book can do so on the MINE platform. In addition, we have included references to our published preliminary work, which in turn each refer to the research literature. At the end of each part, we also provide relevant literature that we consider important on the respective topic.

2.2 Questions are more Important than Answers

This book does not provide *standard solutions* for dealing with environmental problems. Instead, it is an offer on how to create access to complex questions and navigate a path through the thicket of disciplinary perspectives, used concepts, and existing knowledge.

Key Point: The Significance of the Right Question

To effect change, it is crucial to ask **the right questions** at the outset. These questions pave the way towards sustainable development. Experience shows that if the right questions are asked and patiently researched, the answers often emerge over time "as if by themselves". Conversely, solutions offered before a problem has been thoroughly examined often prove to be ineffective in the long run. Hence, we assert: questions are more important than answers.

Serious scientific work primarily involves asking questions. This may seem paradoxical as most people focus on the answers, i.e., the solutions to certain problems and questions. So, why are the questions more important?

Firstly, we must acknowledge that we live in a world where we are constantly surrounded by information. It is available in newspapers, books, news broadcasts, live tickers, on social media platforms, in Wikipedia, scientific studies, or in large data sets, for example, from global internet corporations. The issue is not a lack of information and facts, but rather the overwhelming abundance of it. For instance, during the COVID-19 pandemic, a multitude of key figures, facts, and statistics were readily available and updated daily. Despite this, many people struggled to draw conclusions from the deluge of facts and claims.

The problem is that there is usually an excess of information, but a lack of context to understand what it means—for individuals, societies, or even for specific societal and political action. This context is usually provided by the questions. All information is, in a certain way, answers to the most comprehensive of all questions: What is the case? We are constantly surrounded by a multitude of answers that tell us what is the case somewhere at some time under some circumstances. However, these answers are usually confusing and unsorted, because there is virtually an infinite number of things that are "the case", important and unimportant, related and unrelated. Very little of this has a concrete practical or action-relevant meaning for those who take note of it. Moreover, it is often claimed that something is the case, which turns out to be false upon closer examination (consider *fake news*).

We can generally do little with an answer if we do not know the corresponding question and the appropriate research interest. Accurate information without the corresponding question is in a certain sense a body of knowledge, but it is like a treasure to which we have no access. There is a lack of a structure that makes this knowledge accessible and understandable to us, and there is a lack of a filter that separates the relevant information from less relevant information. The goal of scientific procedure, on the other hand, is to filter information systematically, to gain order, overview, and clarity in the large amount of real and alleged facts. Science, in this sense, is an activity that collects information, structures this information with the help of questions, i.e., organizes it towards a knowledge goal, and in this way transforms a mere collection of information into accessible and usable knowledge.

2.3 The Challenge of an Interdisciplinary Approach

The MINE project and this book take a step further. The following consideration underlies this step: Scientists from individual disciplines collect information in different ways and formulate their insights in different models and theories. However, the terms of one scientific discipline are often not understandable to representatives of another, and one scientific discipline works with assumptions that are seen critically in another discipline. Thus, the connections between different sciences, their intersections and interfaces are often unclear. Even physics and biology have little in common upon closer inspection. It becomes even more difficult when one wants to bring together a sociology shaped by Karl Marx with a modern behavioural biological approach. Communication usually fails here.

Key Point: Interdisciplinarity as the Foundation of this Book
In this book, we propose that the crucial questions must be formulated in a way that **transcends individual scientific disciplines**. We view this contribution as an important step beyond the approach of individual scientific disciplines. It goes without saying that we remain committed to the findings of individual scientific disciplines.

For our book to be read profitably, we expect a certain level of cooperation from the readers. For instance, if we use the same practical example in the chapter on "Responsibility" that also illustrates our chapter on "Stocks", this is not a mere repetition of what has already been discussed. Instead, we aim to convey to the readers the diversity of perspectives from which such a practical example can be viewed. Dealing with sustainability involves understanding, mastering, and independently merging different perspectives. Therefore, this book does not propose an "over-system" or an "over-perspective" beyond physics, economics, sociology, and philosophy. Ideally, the readers will explore the different viewpoints in a sequence of steps that is not mandatory, but represents a well-considered proposal by the team of authors. Ultimately, the readers must independently see how the perspectives come together for them. They need a skill for this, to which we have dedicated a separate chapter. This skill is *power of judgement*. It stands at the interface between knowledge and action, between theory and practice. If one were to summarize the goal of the book briefly, one could say that it is about training the readers' judgement in sustainability issues and providing them with the perspectives that are prerequisites for wise decisions and successful action.

Our considerations do not completely cover the topic of "sustainability". Nevertheless, our selection of concepts can provide orientation and enable a way to deal with the complexity of sustainability issues and environmental problems. The presented concepts can therefore be understood as an "entry" or first step to approach the complexity without causing disorientation. The experience from 40 years of policy advice at the Chair of Economic Theory in Heidelberg shows that significant progress can be made when the complexity is taken seriously. Therefore, we have structured the complexity in the book in a way that we believe presents it accessibly without undue simplification.

There is a certain risk in presenting these concepts here without having people at the readers' side who can facilitate their understanding through conversation, in the exchange of questions and answers, and help them overcome difficulties. We have presented all the concepts dealt with in this book in seminars and other block events—often with several of us teaching together. Particularly fruitful were the discussions with the students, but the discussions among the teachers, who were often able to clarify and formulate their ideas more precisely due to the students' questions, were also stimulating. The fact that several generations have been involved in these ideas and concepts and have shaped their formulation in numerous conversations has certainly also contributed to a clarification and didactic preparation of the questions that concern us. The range becomes clear upon examining the birth dates of the team of authors, which range from 1938 to 1994. We hope that something of the atmosphere of personal exchange is conveyed through the texts of this book.

Our book demonstrates with different examples that it is difficult to act effectively in the face of environmental problems and challenges. However, that is not the main message. Rather, we want to encourage looking at the difficulties, but not being overwhelmed by them. What is needed is the courage to look closely (theory) and to act (practice). It is clear that one cannot see and know everything. Whatever we offer to those who want to do and achieve something with our texts—they remain incomplete hints, they remain written words, and whoever takes it in hand has a book and nothing more than a book in front of them. It is important to us to clarify that this book is not a manual for action, not a recipe for the right sustainability policy. But the book can certainly provide orientation and encouragement. Because our concepts demonstrate that individuals do not need to know everything or possess inherently perfect character traits to implement effective sustainability policies.

Important to know: Demands on good politics

As readers, we wish for committed personalities with civic spirit and political actors who want to take responsibility in social coexistence. This book is intended to help further develop those competencies in them that good politics has always demanded:

- Overview (knowledge) within the realm of the possible, a determined ethical orientation (as directed towards sustainability in this book),
- a sense for the right time to act,
- the courage to act into the open without being sure of the outcome -
- and, in the end, humility, when the actual achievement is never quite and sometimes not even half of what one wanted to achieve at the beginning. Despite all the novelty and complexity of the problem, the challenge of environmental and sustainability policy is essentially the same as that which politics has always faced.

References

Boulding, K. (1966). The economics of the coming spaceship earth. In H. Jarrett (ed.), *Environmental quality in a growing economy,* Resources for the Future/Johns Hopkins University Press.

Krohn, P. (2021, April 18). Nachhaltige Entwicklung: Leben in einer 2-Tonnen-CO2-Welt. *Blog Politische Ökonomie*. https://www.blog-bpoe.com/2021/04/18/krohn/.

Economy, Ecology, and Philosophy

3

Contents

3.1 Changing the World Radically—Immediately and by any Means?

Why do we require philosophy when humanity must fundamentally alter its relationship with its life bases? "The philosophers have only interpreted the world in various ways; the point, however, is to change it." This urgent sentiment was expressed by the young Marx in the eleventh of the *Theses on Feuerbach* from 1845 (Marx, 1969, p. 7). Philosophy, with its diverse interpretations of the world, had its place when there was no immediate action required. However, when action is necessary, deliberating over various interpretations of the world only fosters doubt and delays decisive action.

Marx's words resonate with the zeitgeist of 2023, which is impatiently demanding radical and extensive changes. Climate change, species extinction, rapid loss of natural spaces, declining water reserves, polluted waters, endangered soil fertility, uncontrolled land consumption, plastic waste in the oceans, pollutant particles in the food chain—an almost overwhelming array of indicators show that the life bases and well-being of people are dramatically at risk. Whether one perceives the

M. Faber et al., *Sustainable Action in Economy and Society*,
https://doi.org/10.1007/978-3-662-69122-9_3

world's clock to be at five to twelve or at five past twelve, not only climate and environmental activists, but also many experts are convinced that action is imperative. Hesitation seems inappropriate in the face of impending disasters.

Activists advocating for a turnaround generally have a clear understanding of what needs to be done: Any behaviour that endangers the life bases of humanity must be halted. Whether it's brown coal mining, meat consumption from factory farming, or private car traffic—business as usual is no longer an option, things must change as quickly as possible! However, such a conviction can lead to short-sighted activism. Activism implies that the actors, without considering alternative perspectives, are singularly focused on a specific goal that they prioritise. The appropriateness and efficiency of the means, as well as potential counterproductive consequences of such actions, are often overlooked and not taken into account.

3.2 What One Must Consider for Things to Improve

When attempting to fill the concept of changing today's world with concrete content, it becomes apparent that this is a highly demanding task that can easily lead to overburdening. This task is encapsulated by the term *socio-ecological transformation*. The term *social* addresses the entire field of human relationships in its diversity, while *ecological* refers to the entire area of nature with its abundance of physical and biological processes. The singular term *transformation* can be misleading in this context, as the impending changes involve a multitude of necessary transformations—technologies, consumption habits, institutions, political models, cultural and spiritual aspects are all affected (Manstetten et al., 2021).

All human life is sustained and develops through the interaction of social and ecological factors. For deliberate transformations to yield results that allow people to live a tolerable or even good life based on the natural basis of life, both humans and nature—considered individually in the diversity of their dimensions and together in their complex network of relationships—must be understood appropriately. Conversely, if socio-ecological transformation programs are based on a truncated or flawed understanding of humans and nature, failure is predictable. Well-intentioned actions could even exacerbate the situation.

The desire for change does not absolve one from the responsibility of knowledge. It is necessary to scrutinize: What in society and nature should be changed, and what should be preserved? What promising paths exist, what means should be used, which should not, and which pitfalls must be avoided? What unintended consequences can occur? What must not be overlooked so that changes can occur and lead in the right direction? Are there conflicts of objectives between the two poles of humans and nature: Can measures to preserve the natural basis of life endanger social peace? Can prioritizing social peace lead to the destruction of the natural basis of life continuing unabated?

Questions like these form the background of this book. In conscious contrast to Marx's formulation, one could encapsulate it with the motto: If the world is to be changed, it is crucial to interpret it appropriately. What is the status quo? What happens, how do conditions develop if nothing is done? At which points and to what extent can one change the world if one does something? Is it certain that it will be better if it is changed? What promising courses of action are there?

If the world is interpreted incorrectly or incompletely, then attempts to change it are very likely to fail. Conversely, a realistic, comprehensive interpretation of the world provides a basis for decision-making, with the help of which fruitful changes can be initiated and wrong paths can be avoided. Views on what is, views on what is possible and views on what is necessary, in the realm of the human and the natural,—these are topics that this book addresses. This includes, in particular, being aware of essential structures and processes in nature and society, especially regarding their temporality. Their inherent tendencies, development potentials and possibilities for change must be recognized. This means becoming aware of the respective inertia and immobility, but also the instability, fragility and dynamics of conditions. At the same time, one must always be ready, as environmental and climate activists rightly demand from politics and society, to decisively seize opportunities and time windows for right action (Klauer et al., 2017, pp. 122-139). In all this, however, one must acknowledge one's own ignorance and susceptibility to error (see Chap. 12 *Ignorance*), so that even comprehensive plans will be made with reservations in the face of the possibility of new insights, unforeseeable obstacles or opportunities.

3.3 Environmental Problems and Economics

To this day, there is a tendency in politics, industry and even in the sciences to view environmental problems as tasks for natural scientists and engineers. Natural scientists develop methods to measure the degree of pollution of air, water and soil, while engineers invent technical solutions for the problems identified by scientists. However, this perspective overlooks something crucial: the human factor. The ecological crisis is a result of human behaviour. Therefore, the question of solutions includes questions about the relationship between humans and nature, about the conditions of this relationship and its potential for change.

When we pose these questions, the field of economics comes into view: This includes the areas of production, exchange (e.g. market) and consumption, but also the extraction of raw materials and the disposal of waste materials. The preservation of the natural basis of life depends crucially on how people shape their economy.

The term 'economy' as used here is only partially the subject of economics. Since the emergence of neoclassical economics in the last quarter of the 19th century, many prominent economists have come from the fields of engineering, natural sciences, and mathematics[1]. This led to a formalization of economics that

[1] To name a few: Hermann Heinrich Gossen (1810–1858), Leon Walras (1834–1910), William Stanley Jevons (1835–1882), Marshall (1842–1924), Francis Edgeworth (1845–1926), Vilfredo Pareto (1848–1923), Irving Fisher (1867–1947), Abraham Wald (1902–1950), John von Neumann (1903–1957), Tjalling Koopmans (1910–1985), Gérard Debreu (1921–2004), Kenneth Arrow (1921–2017), John Nash (1928–2015), Reinhard Selten (1930–2017).

often overlooked the human and natural components. Despite significant exceptions[2], conventional economics typically reduces humans to the concept of *Homo Economicus*[3], the selfish rational utility maximizer, whose life orientation is usually focused on personal well-being. This worldview often ignores those aspects of humanity that are crucial for change (see Chap. 5 *Homo Economicus*). Nature is almost entirely disregarded (see, among others, Chap. 9 *Thermodynamics* and Chap. 16 *Foundations of Life*). Nature is only prominently represented in Ecological Economics, established in 1989, which does not belong to conventional economics.

Since the early eighties, a research group, with changing personnel, led by Malte Faber at the University of Heidelberg, has been working on the concepts summarized on the MINE website (www.nature-economy.com). Starting from the belief that the approach of economics is not comprehensive enough, this group has developed a new interdisciplinary understanding of the economy. On the subject side, nature is given a central position; on the methodological side, aspects of social sciences and the humanities, especially philosophy, are explicitly incorporated. The concepts on the MINE website form the basis for the concepts of the book at hand. These are useful for developing an interpretation of the world that provides a basis for decision-making, enabling fruitful changes to be initiated and wrong paths to be avoided.

3.3.1 The Intersection between Natural Sciences and Humanities: Interdisciplinary Economics

This book begins with considerations of what Marx refers to as the *metabolism of man with nature* (Marx, 1962, p. 192). This metabolism occurs within the economy. Raw materials are extracted from nature and transformed via production into goods that serve to satisfy human needs. During and after the use and consumption of these goods, joint products are created (see Chap. 13 *Joint Production*). These joint products often end up as pollutants, exhaust air, wastewater, and waste in the natural environment, contributing to climate change among other issues.

To understand these processes, two aspects need to be considered: one is the human aspect. People make choices about how they live, what they consume, and which profession they want to pursue. They discover raw materials, invent techniques, save to invest capital in innovations, and also develop social, political, and economic institutions by establishing trade routes, using money, setting up

[2]This refers to the field of behavioral economics, established in the last quarter of the previous century, where economists, psychologists, and behavioral scientists collaborate.

[3]Even behavioral economics, established at the end of the 20th century and enriched with methodological elements from psychology and neuroscience, offers only a reductionist understanding of human behavior. However, it does examine scenarios where people act contrary to the Homo Economicus model.

markets, building a legal order, etc. From this aspect, the economy is determined by human ideas about a good life and the ability to technically and practically implement these ideas.

The other aspect is nature. Every economy has a natural basis that significantly determines its possibilities and limits. What enters the economy in terms of material comes from nature, and what leaves it in terms of material goes back into nature (see Chap. 9 *Thermodynamics* and 13 *Joint Production*).

When we examine the economy with its two aspects, we find that it combines perspectives that are usually studied separately: The question of nature leads us to the natural sciences. The question of the human being leads us to the sociological and anthropological[4] research of the social sciences and humanities. The questions about the relationships between humans and nature lead us to a field that combines natural, humanities, and social sciences, which acts as an intersection between the sciences.

Therefore, the metabolism between humans and nature requires many perspectives to be captured in its essential aspects. What is required is both an interdisciplinary scientific view and a philosophical foundation.

Important to Know: Interdisciplinary Economics Acts as an Intersection

Various sciences deal with the two aspects of nature and humans. While processes in nature are studied by the natural sciences, the humanities and social sciences focus on humans. In the middle between the two stands the economy as the metabolism between humans and nature. Interdisciplinary economics can be seen as a kind of intersection between natural and social sciences. On the one hand, it examines the significance of economic production for nature—resource extraction, transformation of resources, and discharge of waste, wastewater, and exhaust air—for which knowledge of natural sciences is required. Furthermore, it examines the needs and interests of people, which determine what is produced, how it is produced, and how the produced goods are distributed (see Chap. 4 *Sustainability*). To answer these questions, knowledge of the social sciences and humanities is required. Thus, insights from both the natural sciences and the social sciences and humanities flow into interdisciplinary economics, influencing each other; hence the image of the intersection. While the discipline-based economics are a suitable starting point for this challenge, it is necessary to expand and deepen the economic perspective in dialogue with insights from the natural, humanities, and social sciences.

[4] Anthropology is the study of humans.

3.3.2 Interdisciplinarity and the Problem of Understanding

This book is intended for individuals seeking an overview of the diverse fields where changes are necessary and possible before taking action. Such an overview allows for the formulation of tasks in all their breadth, depth, and complexity. The individual disciplines of the scientific community cannot provide this comprehensive view of the factors crucial for socio-ecological transformations. Natural scientists typically possess precise and thorough knowledge of their respective areas of nature within their defined research fields. However, even within physics and biology, language and communication barriers exist between the individual disciplines. The language of biology and the language of physics are essentially foreign to each other, often without translation aids. When economic, legal, or social science aspects are introduced, scientific qualifications alone are insufficient for competent participation (see Chap. 12 *Ignorance* and 17 *Summary*). Conversely, expertise in economics and social sciences often lacks depth when it comes to factors from the fields of physics or biology.

Each science has its unique language and methodology. When aspects of different sciences are combined, there is often a Babylonian confusion of terms and methods. The problems do not lessen when the practical knowledge of business economists, engineers, chemists, technicians, and doctors is incorporated. This applied knowledge has the advantage of being effective in reality over theoretical sciences, but it is usually one-sided. The engineer who optimizes engines for electric cars may not necessarily be competent in addressing the ecological, economic, legal, or political problems associated with their ideas. In general, the tasks that emerge with larger changes cannot be solved with the knowledge of a single scientific discipline or in the manner of engineering tasks.

To appropriately formulate the upcoming tasks, communication structures are needed that are open to disciplinary knowledge, but integrate this knowledge into a comprehensive view and are also generally—i.e., not only for the respective experts—accessible and understandable. These structures should enable problems to be viewed beyond disciplinary boundaries in all essential aspects. This is particularly true for human metabolism with nature. Interdisciplinary economics, to be effective as an intersection, relies on such structures. Philosophy plays a crucial role in establishing such communication.

Important to Know: Philosophy as an Approach to Fundamental Questions

This involves a comprehensive view of problems in the human/nature area, which as such belong to no single discipline: These problems are not exclusively physical, biological, economic, political, social, and cultural problems, but they are all of these simultaneously and more. A perspective that accommodates this situation can be described as philosophical—provided that philosophy is understood as dealing with the foundations of our cognition and action.

3.4 Philosophy and Wisdom

Economics has repeatedly been the subject of philosophical considerations[5]. Some of the most significant figures of the so-called economic classics, including Karl Marx, were philosophers by training (Petersen & Faber, 2018). This philosophical view of economics, which has largely been lost to the economic sciences in the last 150 years, needs to be regained in light of upcoming changes. But what does philosophy mean in this context?

Ideally, the major changes that are needed should be based on a comprehensive understanding of human beings and nature, encompassing all essential factors that must be considered by the actors. The ability to comprehend a problem situation in its entirety and to draw the practical consequences necessary for a good outcome is called wisdom in the tradition of Greco-Roman antiquity. The establishment of basic rules for the order of human coexistence was particularly considered a task that could only be solved by a wise personality.

One of the greatest transformations of antiquity was initiated by the Athenian Solon (ca. 640–540 BC), who was the founder of the first constitution with citizen participation. Solon, who laid the foundation for the subsequent flourishing of his city, was considered one of the great sages of Greece. However, it was also known in antiquity that wisdom is often only an ideal: either the conditions to be shaped are too complex, the possibilities of knowledge and intervention of humans are too limited, or the will is lacking to put what is recognized as right into action (Petersen & Faber, 2001).

So, it is not always possible to be wise and to act wisely, so that reasonable goals can be reliably achieved. However, it is always possible to strive for wisdom with all one's might and to act according to the best knowledge and ability. Against the background of such considerations, philosophy has developed. *Sophia* means wisdom, *Philia* is friendship or love. A philosopher is a person who loves wisdom, but at the same time knows that one may never fully attain it.

Those who devote themselves to philosophy, however, take on a central concern from the dimension of wisdom: to comprehend the whole of a situation and, from this overview, where possible and necessary, to derive orientations for action. However, philosophy differs from wisdom not least in that it has developed an independent approach.

---

[5] Notable figures include Aristotle (384/3- 322/1 BC), Jean Jacques Rousseau (1712–1778), Adam Smith (1723–1790), Thomas Robert Malthus (1766–1834), Georg Wilhelm Friedrich Hegel (1770–1831), Mill (1806–1873), and Karl Marx (1818–1883). A concise presentation of the economic ideas of these philosophers, including those of Thomas Hobbes (1588–1679) and John Locke (1632–1704), is given by Petersen and Faber (2018, Part 2).

3.5 The Approach of Philosophy

To this day, we can refer to the four fundamental questions of philosophy as formulated by Immanuel Kant (1968):

1. *What can I know?*
2. *What should I do?*
3. *What may I hope?*
4. *What is man?*

These fundamental questions, which are present in many concepts throughout this book, could also be asked in the field of wisdom. However, philosophy differs from wisdom in how it answers them. In philosophy, one's own ignorance is explicitly addressed (see Chap. 12 *Ignorance*), and it is accepted that within the discourses on theoretical and practical problems, almost every sentence can be opposed by another that must be considered in turn. In philosophy, one becomes explicitly aware of the limitations of human action and the fallibility of the actors, including the individual weakness of the philosophizing person. Nevertheless, philosophy also contemplates (contrary to what Marx claimed in the thesis on Feuerbach cited at the beginning) the question of how far the world should be preserved or changed. *What should I do?,* the second fundamental question of Kant, is the guiding question of practical philosophy, to which this book is committed.

The subject of practical philosophy, insofar as it formulates guidelines for action, is the good life of people in harmony with the course of nature. However, due to the limited possibilities of knowledge and action, a certain modesty always belongs to practical philosophy. Whether a good life of people in complete harmony with nature can be thought of without contradiction under the living conditions of this earth can certainly be questioned. And if such a life were conceivable, to what extent is it realizable and feasible? And, insofar as it would be feasible at least in some aspects, what demands are made on the actors, in what time horizon would it be feasible—immediately, in ten or rather in a hundred years? Or is it already wrong to provide action aimed at transformations with time indices at all (see Chap. 8 *Three Dimensions of Time*)?

Philosophy, as the young Marx suspected, indeed has to do with considerations. But such considerations have nothing to do with hesitation, uncertainty or wavering. They are the considerations that one must have worked through in order not to act blindly. These considerations should not only concern the representatives of the field of philosophy, but above all those who are concerned with far-reaching changes.

Today, it is time for scientists of nature, social and humanities, as well as engineers, business economists, entrepreneurs, but especially civically engaged people, activists to adopt the intention of philosophy to grasp the situation as holistically as possible. At the same time, they should be aware of the limitations of their knowledge and action possibilities and be open to new insights, perspectives and action options in view of the complexity of the problems and the always virulent ignorance.

In the necessary inter- and transdisciplinary dialogue, however, philosophers should also be prominently involved. The chance for real improvements for humans and nature would increase if the power of perseverance, which is nourished by the tradition of the love of wisdom lasting for more than 2000 years, could flow into the program of the upcoming socio-ecological transformations.

In its different concepts, this book tries to correspond to this intention of philosophy, in addition to conveying the respective disciplinary knowledge.

Important to Know: The Significance of Philosophy for Far-reaching Changes

In view of a situation where action is urgently needed, practical philosophy aims on the one hand to gain an overview of the entire field to be worked on, and on the other hand to formulate all knowledge concerning this field in a language developed jointly by all participants beyond the scientific jargons. The four fundamental questions of philosophy should always be present in the background: What can I know, what should I do, what may I hope, what is man? In this context, the ignorance of all participants must always be addressed. Practical philosophy in this sense does not primarily belong to today's specialized discipline that bears the name philosophy, but expresses an attitude that all those interested in preserving the foundations of human life should adopt.

References

Kant, I. (1968). *Akademieausgabe* (Vol. 9). de Gruyter.

Klauer, B., Manstetten, R., Petersen, T., & Schiller, J. (2017). *Sustainability and the art of long-term thinking*. Routledge.

Manstetten, R., Kuhlmann, A., Faber, M. & Frick, M. (2021). Groundwork for Social-Ecological Transformations: The Social Contract, Global Governance and the Meaning of Time, *ZEW* Discussion Paper No. 21-043.

Marx, K. (1962). *Das Kapital. Erster Band*. MEW Bd. 23. Dietz.

Marx, K. (1969). *Thesen über Feuerbach*. MEW Bd. 3. Dietz.

Petersen, T., & Faber, M. (2001). Der Wille zur Nachhaltigkeit. Ist, wo ein Wille ist, auch ein Weg. *Zukunftsverantwortung und Generationensolidarität, Schriften des Institutes für angewandte Ethik, 3*, 47–71.

Petersen, T., & Faber, M. (2018). *Karl Marx und die Philosophie der Wirtschaft: Unbehagen am Kapitalismus und die Macht der Politik*. Karl Alber.

Further Reading: Recommended readings for further study on Part 1 "Starting Points"

Costanza, R. (1992). *Ecological economics: The science and management of sustainability*. Columbia University Press.

Daly, H. E. (1977). *Steady-state economics: The economics of biophysical equilibrium and moral growth*. Freeman. [A classical text of Ecological Economics]

Edenhofer, O., & Jakob, M. (2019). *Klimapolitik: Ziele, Konflikte, Lösungen*. CH Beck.

Gardiner, S. M. (2011). *A perfect moral storm: The ethical tragedy of climate change*. Oxford University Press.

Haddad, B. M., & Solomon, B. D. (2023). *Dictionary of ecological economics: Terms for the new millennium*. Edward Elgar Publishing.

Klauer, B., Manstetten, R., Petersen, T., & Schiller, J. (2016). *Sustainability and the art of long-term thinking*. Routledge.

Krohn, P. (2023). *Ökoliberal: Warum Nachhaltigkeit die Freiheit braucht*. Frankfurter Allgemeine Buch.

Raworth, K. (2017). *Doughnut economics: Seven ways to think like a 21st-century economist*. Chelsea Green Publishing.

Ruth, M. (ed.). (2020). *A Research Agenda for Environmental Economics*. Elgar.

Part II
Humans and Their Actions

4 The Guiding Principle of Sustainable Development and the Concept of Justice

Contents

What's it About?

It is about cake. More precisely, it is about baking and distributing cake, ensuring that there will be cake not only today but also tomorrow, the day after tomorrow, in ten and in a hundred years—for everyone. So it should be just. However, there is a problem: The supply is finite, the replenishment limited. Additionally, the remnants from baking and eating must be cleaned up, otherwise everyone will suffocate in the waste and baking will cease.

This chapter explores how sustainable development shares many similarities with baking and eating cake. The question of the appropriate use of natural resources leads to the question of what is just, how justice is determined in each instance, which needs are necessary for a good life and which are detrimental. We critically examine the notion

M. Faber et al., *Sustainable Action in Economy and Society*, https://doi.org/10.1007/978-3-662-69122-9_4

that constant economic growth is sustainable. Is the idea that the cake must always grow a good one, or does it lead to the long-term result of having no cake at all?

4.1 Introduction to the Concept

4.1.1 The Analogy Between a Cake and the Goal of Sustainable Development

Baking a cake and sharing it with a few friends sounds simple. However, imagine that the cake must be large enough to satisfy not just a few friends, but many people, most of whom we do not know. Now it is no longer so simple: What type of cake should we bake? How big does it need to be? Who should get how much of the cake? The task becomes extremely complicated, however, when we must manage the ingredients in such a way that we can serve a cake to the many guests we expect today, and also tomorrow, the day after, and repeatedly for future guests we do not yet know. Even those who join the party later should receive enough, and in the end, everyone should go home satisfied. But what happens when we find that we have plenty of flour, water, sugar, and eggs for now, but the future supply of ingredients is uncertain? If we bake too large a cake today, the cake of the future may soon be too small given the many people who want to participate in baking and eating cake tomorrow, the day after, and later.

The task of humanity to live sustainably is similar to this cake task. The cake represents the total of goods and services that humanity produces with the resources it extracts from the earth to satisfy its needs. The goal of sustainability also requires a continous supply of ingredients: not merely flour, eggs, and sugar, but all natural resources available today must be used in such a way that the needs of tomorrow and the day after are not neglected. Our guests are all people on earth, now and in the future. Our cake includes the products of the entire world economy. The ingredients, i.e., the natural resources we are using now, must be handled in such a way that future generations also have adequate foundations for their lives. It is not just about meeting basic needs like eating, drinking, and housing. The guests should be satisfied overall, they should be enabled to lead a good life throughout their entire lifetime. It is not enough to serve a tasty cake to those who are here right now, i.e., to secure them an adequate supply of the necessary and pleasant things for life. Rather, one must always have the future guests in mind. Sustainability, therefore, requires careful handling of natural resources.

This requirement is endangered by a central characteristic of human economic activities. The baking and eating of the current cake, i.e., the production and consumption of everything that is available to humanity to satisfy its needs, until today causes massive environmental damage. These damages occur all over the earth in the production, transport, and consumption of goods in the form of waste and emissions. These damages—climate change, loss of biodiversity, water scarcity, destruction of cultivation areas, etc.—can lead to future generations receiving nothing or too little of what they need. In view of such dangers, the following was formulated as early as 1987 in the so-called Brundtland Report:

"Sustainable development is development that meets the needs of the present generation without compromising the ability of future generations to meet their own needs" (WCED, 1987, art. 27).

Sustainable development appears as the guiding principle of the hour or even the century. This guiding principle should provide orientation for politics and society. Politics shall enact laws, commandments, and prohibitions that provide the economy with a framework for action to ensure the long-term survival and the opportunity for a good life for humanity. In society, the guiding principle of sustainability, if not often in deeds, is at least widely accepted in words. In public discourse and at the workplace as well as in the private sphere, there is talk of sustainable lifestyles, sustainable nutrition, of switching to sustainable mobility, of reportable sustainability performances of companies, etc. But do we all mean the same or at least similar things when we talk about sustainability and try to align our lives with it? Is the guiding principle in the eyes of young students in Germany the same as from the perspective of managers of an Indian corporation? While sustainability for some people means a willingness to make far-reaching changes that affect not only politics and economy, but also one's personal lifestyle, for others the term is primarily associated with market positioning, political tactics, or social standing.

There is not a uniform understanding of the guiding principle of sustainable development in our western society, let alone worldwide. But if sustainability is only a shell for arbitrary things and "sustainable" can be printed on every food packaging, every electric car, every company report, and every election poster, then necessary changes can easily be omitted by decision-makers and the foundations of life may continue to be irreparably damaged. The people of tomorrow will be born into a world where they find not even enough to survive, let alone to live a good life.

In the following, we aim to present a universally accessible understanding of the guiding principle of sustainable development. For this, we will

- present the historical development of the guiding principle,
- reflect and discuss the associated standard definition,
- and explore the relationship of the guiding principle of sustainable development to the concept of justice.

We cannot provide a definitive answer here to the question of how we can make our economic system and our way of life sustainable. However, at the end of the chapter, we will pursue a first approximation to an answer by introducing the concept of *sufficiency*.

4.1.2 History of Sustainable Development

Our economic system, which is based on the increasing consumption of natural resources, is causing humans to destroy our environment. This, in turn, endangers the natural basis of life upon which our societies depend.

Certain pre-industrial societies were acutely aware that their survival could be at risk if the natural basis of life were excessively used and unable to regenerate. This could occur through overgrazing, deforestation, or water pollution. However, the environmental problems of these earlier times were local and immediate, and therefore often easily perceptible. Due to their limitations and short-term nature, they could often be solved or circumvented, for example by using new land. In other cases, they led to severe impairments of the affected population or even to the downfall of an entire culture.

It was not until the 1970s that many people became aware that economic activities of humanity, which were largely aimed at improving living conditions, were leading to the exhaustion of necessary resources and causing medium and long-term destruction of the environment on a global scale at an unprecedented intensity. It was understood that if this economic system continued as it had been operating, the living conditions of many, perhaps even most people, would significantly worsen instead of improving. In 1972, the Club of Rome published its report *The Limits to Growth* (1972), which initiated a global discourse on the negative impacts and the finiteness of continuous economic growth. For the first time, a large number of people, including scientists and political decision-makers, expressed the sentiment, "This cannot continue!". This led to the formation of the World Commission on Environment and Development (WCED) in 1983, initiated by the United Nations, which was tasked with outlining a global program for change. The result was the report titled *Our Common Future*[1] (WCED, 1987). This so-called Brundtland Report has shaped the guiding principle of sustainable development for a global public. The formula quoted above has established itself as the global standard definition of sustainable development to this day.

4.1.3 The Essential Concepts in Sustainable Development

The United Nations' guiding principle of sustainable development is based on two central hypotheses:

1. The capacity of individuals to meet present and future needs is limited. This capacity is determined by Earth's resources, including the natural bases of life,

[1] The report is also known as the Brundtland Report, named after the chair of the commission, the former Norwegian Prime Minister Gro Harlem Brundtland.

sunlight and energy, available technologies, political and social organization, and the laws of nature (see Chap. 9 *Thermodynamics* and 11 *Evolution*).
2. Humans of current and future generations have needs that should be met.

The first hypothesis posits that our planet can only provide limited resources and absorb a finite amount of pollutants. For further explanation, refer to the chapters on *Thermodynamics* (Chap. 9) and *Joint Production* (Chap. 13), as well as Rockström et al. (2009) on Planetary Boundaries. The second hypothesis, concerning the needs of present **and** future generations, requires careful consideration. The guiding principle of sustainable development does not specify which needs should be satisfied or to what extent. Basic survival needs such as food, clean drinking water, basic medical care, and shelter are unquestionable, but what about needs beyond these? How should we assess needs for freedom, peace, and solidarity? Countless other needs are conceivable, given the variety of individual, social, and geographical contexts. Depending on the type and extent of the assumed needs, interpretations of sustainability can vary greatly. To put it bluntly, whether sustainability means merely guaranteeing basic needs and thus securing survival as a minimum standard for people, or whether it involves enabling every individual to fulfill all their consumption needs, regardless of whether it is a small car, an SUV, a private jet, or a space flight, makes a significant difference in the possibilities for achieving sustainable development.

We conclude that a standard is needed to determine which *needs* should be met now and in the future, and which should not, in order for humanity's development to be considered sustainable. To identify the needs relevant to sustainability issues, we must ask: Is it just if this or that need, claimed by this or that person or group, is met? Conversely, is the fulfillment of certain needs unjust, perhaps because of the resources required for it, or because of its effects on fellow human beings and creatures? Without the concept of justice, we cannot speak of sustainability, as we have no standard for the needs that must be met—for both present and future generations.

4.2 The Concept of Justice

The everyday meaning of justice can be illustrated using the example of a cake. A cake should be divided in such a way that, as far as possible, the needs of everyone involved—whether as bakers, servers, cleaners, or simply as eaters—are met, without anyone being favored or disadvantaged. Accordingly, when thinking of justice, one initially thinks of the distribution of material wealth, such as property and income: the so-called *distributive justice*. However, this only captures one aspect of justice. Imagine a social order in which everyone is satisfied with their material wealth, but only a portion of the people are free and have the power to dictate what the unfree should think, do, and how they should live. Such a social order would likely appear unjust to most people. So if justice includes essential elements beyond *distributive justice*, what are these elements and how do they fit together?

In the history of philosophy, important thinkers from Plato (428/427-348/347 BC) and Aristotle (380-322 BC) to John Stuart Mill (1806-1873), John Rawls (1921–2022), Martha Nussbaum (*1947), and Amartya Sen (*1933) have endeavoured in various ways to develop an appropriate concept of justice. What these attempts have in common is that they fulfill the demand established in Roman legal thinking, "To each his own"[2]. Justice, therefore, consists of each person having access to the means, circumstances, and opportunities for development that are due to them. This applies on a material as well as on a non-material level.[3] To the non-material rights belong, for example, the human rights of each individual, which guarantee, among other things, freedom and security from encroachment by the state or other individuals and groups.

Such rights, in order to be effectively valid, require both laws in which they are formulated, and authorities that enforce these laws to prevent and, if they occur nevertheless, punish violations. Justice only exists against the background of an order that includes the institutions of law and politics. The state, as the embodiment of these institutions, is required by its constitution to guarantee a legal state, including its continued existence in the future. This includes, in particular, peace within society and peaceful relations with neighbouring states. The stability and continuity of the order are essential elements of justice, and they are also the prerequisites for unjust laws and conditions to be peacefully and orderly changed (Becker et al., 2015). Finally, justice entails that members of a society, on the one hand, adhere to the prevailing laws, meaning they practice legal obedience, and on the other hand, treat other members with solidarity and goodwill. Justice in this comprehensive sense is referred to by the ancient philosopher Aristotle as *general justice*. Since it primarily relates to the legal order and the organization of the lives of members of a society, we will refer to this *general justice* hereafter as *justice as order*. Distributive justice is a sub-aspect of this general justice. It should be noted that only distributive justice directly relates to the satisfaction of needs, while general justice limits and structures the corridor of individual needs to be satisfied.

---

[2] Cicero, *De legibus* 1, 6 19. The classic principle of justice "To each his own" was distorted in its meaning when the National Socialists in 1937 placed it as an inscription over the gate of the Buchenwald concentration camp to deny the camp inmates their humanity. Everyone had nothing but humiliation, mistreatment, torture, and murder to expect as their own.

[3] In this context, the Indian economist Amartya Sen speaks of capabilities. Sen argues that, as far as the natural conditions and potentials of a society allow, every person should receive the resources and opportunities needed to shape their life according to their own conception of happiness (Sen, 2000). Such a society would be just from Sen's perspective.

We conclude: A society can only be just if it fulfills the following six conditions:

- It grants all members a life in freedom based on human rights.
- It creates peace within itself and maintains peaceful relations with all other societies.
- It ensures political and legal stability and is designed for the long term, i.e., continuity.
- It establishes distributive justice by giving everyone what is due to them.
- It enables participation, i.e., the involvement of all members in processes intended to replace unjust laws and conditions with better ones.
- Its members adhere to the law (legal obedience) and practice solidarity and goodwill.

Such a society, however, is nowhere to be found in the world. It is an ideal that present societies should orient themselves towards and which they approximate to varying degrees. In the following, we will explain how the demand for sustainable development derives from general justice. This latter is thus a prerequisite for the former.

Understanding the Various Dimensions of Justice

- **Distributive Justice:** Justice fundamentally involves the distribution of material goods, opportunities, or duties, aiming to allocate or allow each person to receive what is due to them.
- **General Justice or Justice as Order:** For a society to be deemed just, it must ensure individual freedom and maintain peace both internally and externally, beyond the demands of distributive justice. It must also secure a legal order that is adhered to by all (stability). This legal order, which should be changeable in detail, must, however, be designed to endure over the long term in its fundamental aspects (continuity). Furthermore, a just society must enable participation, i.e., involvement in political processes (Faber & Petersen, 2008).
- **General Justice and the Corridor for Need Satisfaction:** General justice creates a corridor of individual need satisfaction. On the bottom, it specifies the lower level of needs to be met, and on the top, it limits individual need satisfaction by considering the needs of all others and the demands arising from the legal order and the stability requirements.

4.3 From the Ideal of a Just Society to the Guiding Principle of Sustainability

4.3.1 Intra- and Intergenerational Justice

Everything discussed so far about justice pertains to the so-called *intragenerational justice*. This refers to the ordering of relationships and the regulation of distribution insofar as a presently living generation of people is concerned. However, the aspects of stability and continuity in general justice imply the guiding principle of sustainable development, because stability and continuity are necessary conditions for a society to persist beyond the present and into the future. To be sustainable, however, a society must achieve more. Stability and continuity, on their own, could also result in maintaining laws and lifestyles that are not sustainable. This cannot happen if society seriously addresses the task arising from the guiding principle of sustainable development: it must also grant future generations the possibility to meet their needs, and at the same time refrain from behaviour that endangers this need satisfaction. In addition to maintaining the rule of law and the transmission and advancement of knowledge and technologies, the primary goal of so-called intergenerational justice is the protection of the natural basis of life and the elimination of anything contributing to their decline. The idea of intergenerational justice extends beyond the demands of general justice through the perspective that the natural environment must be preserved for future generations. This appears as an addition to the perspective of stability and continuity. However, this addition can become problematic when the apparent stability and continuity of a society at a certain point in time is based on a legal order that allows excessive consumption of resources. If this excessive consumption is urtailed abruptly, conflicts, even massive social upheavals are likely to follow, due to the then severely restricted satisfaction of needs of those affected: this is the dilemma we face today.

The demand for intergenerational justice is a central component of the guiding principle of sustainable development and is at the heart of many current environmental debates, especially the climate debate. Nonetheless, to ensure stability and continuity, the needs of the respective current generation must not be overlooked. The guiding principle of sustainable development thus encompasses both intergenerational and intragenerational justice although there can be tensions and conflicts between the two forms of justice. For example, in the terms of intragenerational justice, if all owners of second cars were ompelled to transfer them to people without cars, the environment would likely be burdened more than before due to the then increased use of these cars as primary vehicles, resulting in higher fuel consumption and emissions. This would not be conducive to intergenerational justice.

Similarly, a societal debate overly focused on intergenerational justice can lead to the needs of currently living people not being adequately considered. For example, if overly ambitious measures in the sense of climate protection lead to such a sharp increase in energy prices that large sectors of industry can no longer maintain their production, this can result in unemployment and loss of prosperity, even leading to poverty. Furthermore, the consequence would be that mobility would be massively restricted and that many homes would no longer be adequately heated. In connection with this, social unrest could break out, which in extreme cases could even lead to civil wars.

To determine what level of need satisfaction might be appropriate to ensure both intra- and intergenerational justice, a consideration of the third condition of general justice is necessary: the attainment of what is due to a person.

4.3.2 What is an Individual Entitled to? The Realm of Possibilities for Needs

A first answer to what is an individual entitled to and therefore should receive is:

> *In principle, every individual is entitled to everything that enables them to lead a good life.*

However, this answer is not very meaningful, as what appears to an individual as a good life depends on the individual's personal, social, cultural, and geographical context. Fundamentally, every individual has the right to have their chosen needs met within their means, as long as no harm is done to anyone presently or in the future. But this choice is not arbitrary:

1. On the one hand, there is a minimum requirement of what a person is entitled to, without having to choose it specifically. These are the basic needs, without the fulfillment of which a good life is not possible.
2. On the other hand, there are constraints on the fulfillment of needs that an individual can strive for, due to the demands of general justice.

A necessary condition for being able to lead a good life is the fulfilment of basic needs. People who are starving, people who can only inadequately protect themselves from cold and heat, people who have no access to clean drinking water or basic medical care, are so severely restricted that they are not able to lead a life that appears to them as a good life. Basic needs must also include a minimum level of access to education—such as the opportunity to learn reading, writinging and arithmetic—because living well in today's world without access to education is hardly possible. It is therefore the task of any kind of sustainable development to ensure the fulfilment of basic needs, today and in the future. These provide a minimum of what every human being on earth is entitled to.

The realm of possibilities of what an individual is entitled to is limited by the demands and restrictions that the general public imposes on the individual on the basis of the guiding principle of sustainable development. Because the fulfilment of many material needs is associated with high depletion of resources and/or a burden on the environment. Since this endangers the well-being of current and future generations, society must limit or completely prohibit the satisfaction of such needs. Only in this way can it protect its own stability and continuity. For example, while an individual may desire regular consumption of a beef steak, society has reasons to significantly restrict such a habit due to the climate-damaging nature of this behaviour.

We summarize: In principle, a person is entitled to everything that gives them the opportunity to lead a good life. For this, the fulfilment of basic needs is absolutely necessary. What a person is entitled to is limited by restrictions that are justified in the endangerment of current and future generations. However, the restrictions must not go so far that only basic needs appear legitimate. In addition to human rights and the other requirements of general justice (see above), good educational opportunities, intact social relationships, good working conditions, the containment of social inequality, and opportunities for recreation in an intact nature are relevant in today's world.

Important to Know: Justice and Sustainability

- **Sustainability and Intragenerational Justice:** A humanity that follows the model of sustainability requires general justice and especially distributive justice. This means that one goal of sustainable development is to create intragenerational justice, i.e., justice between the different members of the currently living generation. This implies, for example, that a balance must be achieved between the rich countries of the global North and the poorer countries of the global South. Also, the current extreme unequal distribution of income and wealth in many countries and especially worldwide cannot, because it is unjust, be sustainable.
- **Sustainability and Intergenerational Justice:** The model of sustainability requires that current generations deal justly with the life possibilities of future generations, i.e., that intergenerational justice is established. For example, current generations have the task of taking measures against the pressing global problems of climate change and species extinction so that later generations can still live well on Earth.
- **The Corridor of Needs for a Good Life:** The needs of individuals or groups can develop freely within a certain corridor, which, however, must meet certain minimum conditions and is limited upwards by the demands of intra- and intergenerational justice. The minimum condition is set by the basic needs, without the fulfilment of which a good life is not possible. The restrictions of the corridor result from the demands of the general public (including future generations): It must be prevented that excessive satisfaction of some people's needs takes away the opportunity for other people to lead a good life.

The standard definition of sustainable development introduced at the beginning, as development that meets the needs of the current generation without endangering the ability of future generations to satisfy their own needs, becomes more tangible in connection with the concepts of intra- and intergenerational justice. The first step, the creation of a fundamental understanding of sustainability, has thus been taken.

The second step is to consider what the path to a sustainable society might look like. In the following sections, we show that sustainability cannot be achieved by simply producing more and more. The idea that a world in which more and more is produced and consumed is sustainable is fed by the so-called growth paradigm. We point out deficiencies in this paradigm and instead refer to alternative approaches.

4.4 Sustainability and the Growth Paradigm

For many decades, the prevalent belief in society has been that sustainable development can only be achieved through continuous economic growth. The idea is that economic growth can ensure the fulfilment of the basic needs of all people, particularly in countries where many still live in poverty, hunger, and without medical care. Economic growth enlarges the cake for everyone, providing more to distribute. If this surplus were directed towards the poorer people, the wealthy would not even have to relinquish any of their wealth, and conflicts arising from a redistribution of wealth within countries and from countries of the global North to countries of the global South could be avoided. The notion that growth is the solution to almost all societal problems is referred to as the growth paradigm. Steady growth appears to guarantee stability and continuity of societies worldwide. However, this is a fallacy: Although in a society like China, continuous economic growth over the past decades has significantly improved the living conditions of most people. It is foreseeable that many people are already suffering from the environmental problems that accompany growth—such as water shortages, droughts, and floods due to climate change—and that in the future, many more people will suffer to a much greater extent. Therefore, since the Club of Rome report "The Limits to Growth" was published in 1972, the growth paradigm has been in crisis. This crisis is accompanied by the realization that the previous growth of the world economy has not prevented the gap between rich and poor from widening in almost all societies. The poorest people worldwide are largely or completely excluded from the benefits of growth, while the rich and super-rich often benefit disproportionately. This contradicts any reasonable idea of distributive justice.

In response to criticism of the growth paradigm, the concept of "green growth" was developed. Its proponents do not see growth as a problem, on the contrary: "Environmentally friendly growth" is, according to their conviction, entirely in line with the guiding principle of sustainable development. Their argument is: growth, as in the past, will lead to technical innovations and increasing efficiency in the future. This could reduce excessive resource consumption and negative environmental impacts of the economy and almost completely avoid them in the long term. However, this belief in growth does not sufficiently take into account the fact that even the most effective technologies are subject to physical limits (see Chap. 9 *Thermodynamics* and 13 *Joint Production*). Moreover, efficiency gains that could benefit the environment are often "eaten up" by increasing production (see also literature on the rebound effect[4]). For example, converting an irrigation system in agriculture to efficient drip irrigation does not necessarily alleviate the problem of water scarcity, but sometimes even exacerbates it (Grafton et al., 2018). This does not mean that there should or can be no more growth and development. Growth and development will continue to exist, and even need to exist, in many areas. However, in an economy oriented towards sustainable development, growth should not be an end in itself, but a carefully used means. The purpose of such an economy would be to create the possibilities for as many people as possible, today and in the future, to lead a good life.

Some economic sectors, especially in the social area, would grow, while others would be curtailed. An attempt to provide guidance for an economy oriented towards the good life of people beyond the growth paradigm are the 17 Sustainable Development Goals (SDGs), which were issued by the United Nations for the period up to 2030 on January 1, 2016.[5] Some of these goals directly relate to the economy and its growth. A larger number of goals are social in nature, such as reducing poverty, strengthening equality, or access to good education. In this way, particular account is also taken of those human needs that go beyond material prosperity.

The preservation of natural life-support systems is often viewed as a problem in some goals, but it does not receive the importance that corresponds to its significance. These life-support systems are the crucial prerequisite for further life on Earth. The fact that the pursuit of one goal can conflict with the pursuit of another sustainability goal is not sufficiently addressed in the SDGs. However, it is at least recognizable that an understanding of sustainability, which extends beyond material needs, can lead to political objectives—in this case, set by the United Nations—focusing more on the creation of justice in a comprehensive sense and thus enabling a good life for all.

[4] (Alcott, 2005; Jevons, 1865). The rebound effect is referred to when the potential savings from efficiency improvements are not or only partially realized. The phenomenon is also known as *Jevons' Paradox*, in reference to its first description by the British economist William Stanley Jevons (1835–1882). Jevons recognized that more efficient use of coal paradoxically did not lead to the desired savings of the resource, but rather to the opposite: an increase in coal consumption.

[5] The goals are named as follows: 1) No poverty; 2) Zero hunger; 3) Good health and well-being; 4) Quality education; 5) Gender equality; 6) Clean water and sanitation; 7) Affordable and clean energy; 8) Decent work and economic growth; 9) Industry, innovation, and infrastructure; 10) Reduced inequalities; 11) Sustainable cities and communities; 12) Responsible consumption and production; 13) Climate action; 14) Life below water; 15) Life on land; 16) Peace, justice, and strong institutions; 17) Partnerships for the goals. (see United Nations (2016), https://sdgs.un.org/goals).

4.4.1 Sustainable Action and Ignorance

Determining whether a specific policy, economic sector, or behavior is sustainable is often difficult because sustainability issues are highly complex and not easily simplified. This complexity arises from the long timeframes that need to be considered, from the often insufficient understanding of physical and ecological interrelationships, and finally from the diversity of the respective societal and economic prerequisites and possibilities worldwide—for which the climate problem is just one example. Due to this complexity, we have to deal with *ignorance*, which is often not reducible (see Chap. 12 on *Ignorance*).

Nevertheless, it is necessary that statements about the sustainability of actions, policy decisions, or economic sectors are made, because a reasonable choice has to be made among the multitude of problem descriptions and solution possibilities offered. Almost all of the concepts presented in this book contain hints on what needs to be considered in such a choice. Thermodynamic relationships and joint production (see Chap. 9 and 13) play a role, as do temporal structures of natural and societal processes and the organization of responsibility and the understanding of political action (see Chap. 5 on *Concept of Man* and 7 on *Power of Judgement*). The explanations of this chapter therefore offer no more than initial guidance. Based on the guiding principle of sustainable development, which extends to the needs of people both in the present and in the future, our considerations on justice provide a framework that illustrates the criteria under which the realm of possibilities for each individual's pursuit of a good life should be structured and limited. Another orientation is given by the task of preserving the natural life-support systems (see Chap. 15 on *Stocks*), which has absolute priority in sustainability issues in the medium and long term. A loss of life-support systems endangers both the stability, the continuity, and the possibilities for a good life, in the present and in the future.

With this guidance and especially with the concrete hints in the mentioned and other chapters, sustainability issues can be theoretically clarified and practically addressed in the diversity of their aspects. Due to the *ignorance*, there is usually no optimal ("most sustainable") option. But there are good options and less good ones. It is already a significant step towards sustainability when it is clearly recognized which paths should not be chosen. The actors are dependent on their *power of judgement* in their decisions (see Chap. 7). Therefore, people who are challenged to make decisions about sustainability need not despair at the difficulty of the question. Instead of despairing and leaving the field to others, less trained, they can make important contributions to a sustainable society.

4.4.2 Approaches to Sustainable Development Beyond the Growth Paradigm

The label "sustainability" often proves to be beneficial today for the marketing of products, the re-election of politicians, or even the social reputation in a group. The so-called "greenwashing", i.e., the unjustified labelling of an action or product as sustainable, thus becomes a societal problem. Through careful scrutiny and an awareness of the meaning of sustainability, greenwashing can be recognized for what it is. Based on the considerations of this chapter, however, it remains uncertain how a more sustainable world can be achieved. In conclusion, however, we will outline an approach that we consider promising: A *politics of sufficiency,* which includes the good life as a goal and thus shows an alternative to the growth paradigm.

The concept of sufficiency involves determining the appropriate level of consumption (Faber & Manstetten, 2014, Chap. 12; Schneidewind & Zahrnt, 2013). In other words, how much do we need for a good life? Societies that pose these questions take a different path towards sustainability than the one typically pursued. Until now, sustainability policy has often been dominated by the question: "How can we meet the already enormous and still growing demand for goods due to population growth without destroying our natural life support systems?" The answer usually involves a demand for more growth, which should become as "green" as possible through the use of modern technologies and continuous innovation. A policy of sufficiency offers an alternative. It does not exclude the use of modern technologies. However, instead of making significant—and often ultimately futile (cf. Chap. 9 *Thermodynamics* and 13 *Joint Production*)—efforts to create the supply that can meet a continuously growing demand, the consideration of what is really needed can limit the demand to an appropriate level. In parts of the energy industry, this approach is aptly summed up with the phrase: "The cheapest and most climate-friendly kilowatt hour is the one that is never consumed." A policy of sufficiency, like the usual approaches, still aims to produce electricity or other goods to meet a demand. However, it is a demand that is guided by a measure derived from the requirements of a good life. Sufficiency, therefore, should not be equated with a demand for renunciation. For people who already consume excessively, this may be the case. However, a policy of sufficiency also implies enabling those people who live in poverty and whose basic needs are not met to have a higher level of consumption, which can even increase environmental pollution.[6] To meet the ongoing but significantly reduced demand in certain areas in an environmentally friendly manner, a society oriented towards sufficiency also

[6]This increase could, however, be at least partially compensated by a decrease in the consumption of wealthier people. Currently, the 10% highest income people worldwide are responsible for about 50% of global CO_2 emissions (Kartha et al., 2020). The lower 50% of humanity (measured by income) in turn only cause 7% of global emissions.

relies on modern environmentally friendly technologies such as renewable energy generation. The principles of a policy of sufficiency are:

1. Priority is given to determining demand through an inquiry into the appropriate level of consumption.
2. Secondarily, the aim is to generate the supply for the remaining demand using modern technologies in a resource-saving and environmentally friendly way.

The term 'policy of sufficiency' primarily does not refer to large-scale politics, which can only achieve sustainability within its means, i.e., through commands, prohibitions, or structural measures. Since sufficiency concerns the good life, which ultimately arises from the free choice of individuals within the framework of justice considerations, it cannot be mandated by political institutions. A policy based on the idea of sufficiency rather requires the insight that individual approaches to sustainability can be of great importance. Thus, many approaches to a sustainable lifestyle start with individuals. Over time, they can be adopted by larger social groups. An example of this is the increasingly widespread idea in society of limiting one's private life to a comparatively small living space within a community, while at the same time experiencing the value of communally used areas as an essential ontribution to a good life.

However, the success of individual changes is limited, as individual actions are also embedded in social and institutional contexts. Thus, communal living projects can only be implemented with difficulty if funding is lacking and building plots are awarded to more lucrative, but less sustainable individual property projects. In addition, many people feel overwhelmed by the constant task of individually deciding what is the right measure in terms of sustainability. Therefore, in the context of sufficiency, there is indeed an important task for "big politics". It is up to them to create institutional conditions for a sustainable lifestyle. This means that individuals and groups, but also associations and companies, who independently strive for sustainable lifestyles, are encouraged and promoted. At the same time, a policy of sufficiency cannot avoid the task of limiting life habits that obviously contradict sustainable development.

Against this, one will still hear the argument that individual need satisfaction is an entirely private matter that politics should not interfere with. We have shown that aspects of general justice and especially intergenerational justice deprive this argument of its basis. If the natural life support systems are endangered by the behavior of individuals, society has the right and the duty to restrict or prevent such behavior in order to ensure its stability and continuity as well as the good life of all in the long term.

References[7]

Alcott, B. (2005). Jevons' paradox. *Ecological Economics, 54*(1), 9–21. https://doi.org/10.1016/j.ecolecon.2005.03.020

Becker, C., Ewringmann, D., Faber, M., Petersen, T., & Zahrnt, A. (2015). Endangering the natural basis of life is unjust: On the status and future of the sustainability discourse. Ethics, Policy & Environment, 18(1), 60–67.

Faber, M., & Manstetten, R. (2014). *Was ist Wirtschaft?: Von der politischen Ökonomie zur ökologischen Ökonomie*. Karl Alber.

Faber, M., & Petersen, T. (2008). Gerechtigkeit und Marktwirtschaft–das Problem der Arbeitslosigkeit. *Perspektiven der Wirtschaftspolitik, 9*(4), 405–423.

Grafton, R. Q., Williams, J., Perry, C. J., Molle, F., Ringler, C., Steduto, P., Udall, B., Wheeler, S. A., Wang, Y., Garrick, D., & Allen, R. G. (2018). The paradox of irrigation efficiency. *Science, 361*(6404), 748–750. https://doi.org/10.1126/science.aat9314.

Jevons, W. S. (1865). *The coal question: An inquiry concerning the progress of the nation, and the probable exhaustion of our coal-mines*. Augustus M. Kelley.

Kartha, S., Kemp-Benedict, E., Ghosh, E., Nazareth, A., & Gore, T. (2020). *The carbon inequality era: An assessment of the global distribution of consumption emissions among individuals from 1990 to 2015 and beyond*. Oxfam, Stockholm Environment Institute. https://doi.org/10.21201/2020.6492.

Rockström, J., Steffen, W., Noone, K., Persson, Å., Chapin, F. S., Lambin, E. F., Lenton, T. M., Scheffer, M., Folke, C., Schellnhuber, H. J., Nykvist, B., de Wit, C. A., Hughes, T., van der Leeuw, S., Rodhe, H., Sörlin, S., Snyder, P. K., Costanza, R., Svedin, U., & Foley, J. A. (2009). A safe operating space for humanity. *Nature, 461*(7263), 472–475. https://doi.org/10.1038/461472a.

Schneidewind, U., & Zahrnt, A. (2013). *Damit gutes Leben einfacher wird: Perspektiven einer Suffizienzpolitik*. Oekom.

Sen, A. (1999). *Development as Freedom*. Oxford University Press.

WCED. (1987). *Our common future* (Brundtland Report). World Comission on Environment and Development.

[7]The contents of this concept are based on: Faber, M., Frick, M., Zahrnt, D. (2019) MINE Website, Sustainability & Justice, www.nature-economy.com

Homo Economicus and Homo Politicus: How we See Ourselves as Humans and Why that Can Change Everything

5

Contents

What's it About?

This text explores concepts of man, focusing on what it means to be human. Although a clear definition of being human is elusive, certain dimensions such as biological, economic, cultural, and political can be discerned in different concepts of man. These concepts are also utilised in scientific disciplines as simplifications for analytical purposes.

The discussion centres on two specific concepts of man and their implications for addressing sustainability issues. Traditional economics primarily uses the concept of the economic human *(Homo Economicus),* which portrays humans as *selfish rational utility maximizers*. This image is crucial for practical sustainability policy as it forms the basis for economic and environmental policy instruments. However, significant transformations require more than rational egoism from the actors. The existence of individuals committed to their fellow humans and the

M. Faber et al., *Sustainable Action in Economy and Society*,
https://doi.org/10.1007/978-3-662-69122-9_5

long-term preservation of our livelihoods suggests an intrinsic interest in justice and the ability to articulate it in public discourse. This aspect of being human is subsequently described in the concept of the political human *(Homo Politicus)*.

5.1 Introduction to the Concept

The question: What is man? is the starting point of anthropology (from Greek. *anthropos*—man—and *logos*—concept), the study of man and humanity. Anthropology seeks universally valid determinations of being human, yet being human as such defies any definition. Consequently, there is not *one* universally valid anthropology, but rather *anthropologies* from different perspectives. By abstracting from individual human beings and their unique characteristics, these anthropologies develop universally valid ideas of being human, known as concepts of man, within their respective perspectives. Concepts of man are constructions that highlight certain aspects of human nature beyond individuality. A person who fully corresponds to such a concept of man in reality is *not* to be found. However, under certain circumstances, the behaviour of many people may show lesser or greater similarities to the behaviour ideally assumed in a concept of man.

Since many perspectives on man are possible, each concept of man refers to a limited aspect of being human. Within a scientific discipline, the respective concept of man represents a methodically constructed reduction that makes the complex subject of investigation "man" accessible for the disciplinary inquiry. Thus, economics has developed an image of the economic human, the Homo Economicus.

5.2 The Concept of Man of the Homo Economicus

To scientifically explain economic processes and model them mathematically, economists base their models on certain assumptions about the behaviour of economic humans. John Stuart Mill (1806–1873), from whose ideas the concept of man of the Homo Economicus emerged, writes: "Political Economy considers mankind as occupied solely in acquiring and consuming wealth; and aims at showing what is the course of action into which mankind, living in a state of society, would be impelled if that motive, except in the degree in which it is checked by the two perpetual counter-motives above adverted to, were absolute ruler of all their actions. Not that any political economist was ever so absurd as to suppose that mankind are really thus constituted, but this is the mode in which science must necessarily proceed." (Mill, 1974, p. 901f, quoted from Manstetten, 2000, p. 48). Such considerations were the basis for the so-called Homo Economicus approach of traditional economics. The human being is seen as a selfish, rational utility maximizer (Müller, 1989). The Homo Economicus approach contains basic assumptions about how people decide about their need satisfaction and the principles they use to choose the means to achieve it.

Essential for the behaviour of Homo Economicus is the orientation towards self-interest. She knows her own need structure (economically: her preference order), the income available to her and all possible options for action. Due to her rationality, she is always able to optimize, i.e., to make the optimal decision for herself. The criterion for every decision is the optimal satisfaction of her needs.

In the standard economic models, the Homo Economicus is assumed to be 'insatiable'. This means: If there is a better alternative to a (goods) choice (i.e., more goods for the same budget), then Homo Economicus always chooses "more goods". She always wants to have more than she actually has, regardless of how much she already has. It is further assumed that the well-being of Homo Economicus is not influenced by that of others. Envy and pity are alien to her. In economics, this assumption is called *independence of preferences*. The actions of one person are not influenced by others and they do not influence the needs of others.[1] Insatiability and independence must usually be assumed for the calculation of economic decisions.

Important to Know: Assumptions about Homo Economicus

- **Optimal Satisfaction of Needs:** The criterion for economic decision-making is the optimal fulfilment of needs (under given conditions, e.g., available income).
- **Insatiability:** Homo Economicus always wants to have more, at least of one good, regardless of how much she already has.
- **Unlimited Preferences:** All types of needs and interests are allowed. Neither the state, society, tradition, nor religion influence economic decisions.
- **Independence of Preferences:** Homo Economicus cares exclusively about her own well-being. She knows neither envy nor pity.
- **Perfect Information:** Homo Economicus knows her own need structure (individual preference order) as well as her income, all prices for goods relevant to her, and all given action options.
- **Consumer Sovereignty:** Individuals are trusted to have enough insight and judgement to adequately take care of their own needs and interests.

5.2.1 What Does the Concept of Man of Homo Economicus Achieve?

The Homo Economicus assumption is a construction of economic sciences, but it has a concrete reference to reality. It reminds us that in reality, many people prioritize their own needs and advantages over the interests of fellow human beings

[1] Moreover, it is assumed that the Homo Economicus conceals her true preferences if she expects to gain an advantage from it. This concealment is so prevalent that certain economic publications refer to an "incentive to lie".

and creatures. It makes it understandable that many people change their behaviour more due to the threat of financial losses or penalties than due to compelling moral sermons. From the perspective of Homo Economicus, it is also understandable that people, enticed by the prospect of profit, forget all their moral principles. If such widely known facts are duly considered, they can provide meaningful explanations for economic and societal processes. Many environmental policy recommendations are based on insights from economic sciences: this applies, for example, to charges on wastewater and certificates for greenhouse gases.

The fact that financial incentives, which appeal to self-interest, can be used as a tool of politics, shows that the concept of Homo Economicus is relevant for environmental and sustainability policy. Any environmental and sustainability policy that does not take into account Homo Economicus in far-reaching and profound changes is doomed to fail. Resistance, such as that arising from self-interest, rejection of change, and lack of willingness to sacrifice, must be anticipated by political actors in advance to avoid unpleasant surprises when realizing ideas for comprehensive transformations. The concept of Homo Economicus has proven itself in economic sciences, especially in the analysis of standard situations across various markets. But even outside of markets, one must be prepared to encounter behaviour patterns that correspond to the concept of Homo Economicus.

5.2.2 Problems Arising from the Concept of Man of Homo Economicus

The perspectival nature of Homo Economicus is often overlooked, even by economists (Becker et al., 1996). Consequently, this concept can become autonomous and be regarded as the perspective on humanity that is closest to reality. The rational egoist then appears as the norm, or even as a kind of standard in the life-world. Anyone who does not conform to this norm, who does not use "elbows" in competition to push others away, is considered "at fault" if they are not successful. Accordingly, there are designs of human coexistence based solely on the economic concept of man.[2] From the descriptive statement: "This is how—in the manner of Homo Economicus—man behaves under certain circumstances," a value statement is derived: "It would be good and sensible if people behaved this way—in the manner of Homo Economicus—under all circumstances."

The view of economic man as a self-centered, insatiable utility maximizer has repeatedly provoked criticism. The Homo Economicus is criticized for lacking aspects of justice, which are of considerable importance for action towards sustainability (see Chap. 4 *Sustainability*). The criticism begins with egoism and the resulting limitations. Apparently, the Homo Economicus knows no motivation

[2] Here, the New Political Economy (Public Choice) is particularly noteworthy, which aims to provide an independent analysis of political structures based solely on the assumption of Homo Economicus (Buchanan & Tullock, 1962).

beyond his private preferences and dislikes. Impartial interest in a matter or ethical drives that let him share in the well-being and suffering of other people and other living beings only occur to him by chance and in exceptional cases. If the general conviction spreads that people are there solely to care for their own well-being, not for the well-being of the community or even of humanity and the earth, this leads to the potential paralysis of many initiatives aimed at change. This can pose a real danger to successful sustainability policy. If only selfish motivations are assumed to be the very impulses of all people, this leads to a fundamental mistrust of all those who stand up for justice and the protection of nature. The assumption that all those who want to change something are in truth only pursuing private interests is then accompanied by a justification of one's own inactivity and resignation—"because people are just like that." Decisions oriented solely towards egoism are not compatible with sustainable consumption patterns and social solidarity. However, we aim to demonstrate in the following: people are not like that—at least not completely.

Important to Know: Achievements and Limits of Homo Economicus

The concept of man as Homo Economicus reminds us of the power of private self-interest. It describes a dimension of being human that must always be reckoned with in practice. It also expresses a certain trust in individual freedom: Without considering Homo Economicus, sustainability policy cannot succeed. On the other hand, if all people behaved exclusively like Homo Economicus under all circumstances, there would be no assumption of social and ecological responsibility. Especially when responsibility extends beyond one's own life horizon and refers to future generations, a Homo Economicus will evade this responsibility. But since there are people who take on social and ecological responsibility, a concept of man must be conceived that takes this fact into account.

5.3 The Concept of Man of Homo Politicus

In reality, people do not always correspond to the concept of Homo Economicus. Accordingly, it makes sense to operate with concepts man other than those of economics outside the field of economics. Within the framework of the topic of our book, the field of politics is of particular interest. On the one hand, there are processes in politics that can indeed be described with the concept of man as Homo Economicus. These are the processes in which the greed for power and the private advantage of political actors cause either proven structures to be destroyed or necessary changes to be omitted. But where political action leads to substantial improvements, an analysis that only uses the approach of Homo Economicus will fall short. Especially in political action, people repeatedly show that they act from motives that cannot be traced back to their private self-interest. The civil rights movement in the USA in the 1960s can be cited as a historical example of this. In the Federal Republic of Germany, the green movement has owed its greatest successes since the 1970s to the commitment of countless citizens who campaigned for the preservation of nature—although most of them did not derive any private advantage from it. In general, many people are committed to the common good beyond the realization of their own interests. This commitment reveals a dimension of being human that cannot be described with the concept of man as Homo Economicus. The motivations, orientations and characteristics of action that are important for political action in general and especially for action towards sustainability can be formulated with the concept of man as Homo Politicus.

Important to Know: Five Characteristics of Homo Politicus (Petersen et al., 2000)

- **Intrinsic Interest in Justice:** Homo Politicus has an intrinsic interest in the common good and strives for justice.
- **Interest in a Culture of Debate and the Public Sphere:** Homo Politicus is capable and willing to publicly dispute with others and engage in debates about what is just in a particular situation. She wants to convince others, but is also willing to be convinced by others through better argumentation.
- **Courage:** Homo Politicus is willing to accept personal disadvantages for her commitment to the common good and justice.
- **Power of Judgement and Creativity**: Homo Politicus can assess the special circumstances of a situation and recognize the right time to act. She is capable of initiating new and unexpected developments.
- **Sense of Responsibility:** Homo Politicus takes responsibility for the consequences of her actions, even for side effects that were unpredictable in advance.

We will now explain these five aspects:

On 1: *Intrinsic interest in justice*: Homo Politicus strives for the common good. The reason for his political engagement lies in his intrinsic interest in justice. This interest does not necessarily contradict self-interest, but it causes self-interest to be potentially subordinated in favor of the interest in the common good. For Homo Politicus, a decision is just if all those affected can agree to it with good reasons (ideally). However, this definition is associated with a problem: What constitutes good reasons and what does not, often cannot be determined unequivocally and uncontroversially. Consequently, the concept of justice is *not operationalizable.* Another problem arises: Decisions of sustainability policy always also affect future generations. However, one can never be certain about their possible consent, but can only make plausible assumptions. How Homo Politicus deals with such difficulties is addressed in other sections. (see Chap. 4 *Sustainability*).

On 2: *Interest in argument culture and public sphere*: Homo Politicus does not define what is just on their own. Instead, he acts in the public sphere and seeks political debate with others who are also interested in public affairs. To achieve his goal—the consent of the participants to his ideas of justice, who often have very different interests—Homo Politicus must be able to understand the interests and perspectives of the participants. This ability is a prerequisite for Homo Politicus to act in a way that can potentially find the consent of the participants. The process of debate requires freedom of speech and public appearance. In this process, Homo Politicus engages in disputes with those who think differently, tries to convince them and is willing to be convinced by the better argument. Homo Politicus is interested not only in the result of the political process, but in the process itself. For Homo Politicus, competition in the political debate is not an obstacle, but a stimulus.

On 3: *Courage*: With the public presentation of his conviction and in pursuit of his goal, i.e., in the effort to practically realize his convictions, Homo Politicus exposes himself and thus exposes himself to criticism, including unfair and irrelevant criticism. He often finds himself confronted with a media campaign against him or even a *shitstorm* (see Krohn, 2023, Chap. 7 *Power of Judgement*). In view of this, Homo Politicus needs the courage to publicly stand up for his convictions. He must publicly face the attacks against his person and be willing to accept the associated risks and disadvantages.

On 4: *Power of Judgement and creativity*: The Homo Politicus does not merely implement the recommendations of scientific experts. Instead, he utilises a unique *know-how*, a knowledge of *how to do it*, to adapt and implement these recommendations according to the given circumstances. This results in practical solutions that are suitable for the respective political, economic, and social conditions. We refer to this know-how as power of judgement (see Chap. 7). Political theorist Hannah Arendt emphasises that politics as "action" can contain a spontaneous element and can produce something new (Arendt, 1981, p. 166). This connection between politics and creativity aligns with our everyday understanding. We view politics as the "art of the possible". In politics, there is always something to shape. Therefore, we assume that the Homo Politicus's individual actions always have an element of unpredictability, as he has the potential to bring forth something new. Because politics can change laws, institutional conditions of the economy and society, as well as norms and preferences, it represents a possible source of societal change.

On 5: *Sense of Responsibility:* The Homo Politicus, despite not being able to predict all the consequences of his actions, takes responsibility for these consequences. This means that he accounts for the motives of his actions, reviews all the results of his actions, including the unintended ones, and willingly informs the public about his intentions and the actual outcomes. He is also willing to take responsibility for mistakes and omissions (see Chap. 6 *Responsibility*).

It is important to note that the actions of Homo Politicus should not automatically be considered beneficial or useful for the common good. Especially when political actors lack even one of the five mentioned characteristics of the Homo Politicus, their actions can become problematic or even dangerous. A lack of power judgement, a lack of a sense of responsibility, or a rigid concept of justice that neglects the ability to reach consensus within the political community can lead to individuals who consider themselves Homines politici causing significant damage to their society.

5.4 Homo Economicus, Homo Politicus and the Liberal Constitutional State

Homo Economicus requires the liberal constitutional state, with its commands and prohibitions, to pursue his private needs as undisturbed as possible. Threats of punishment deter others from causing him harm and prevent him from choosing paths of utility maximisation outside the realm of legality. Homo Economicus is not inherently interested in the constitutional state, but relies on its existence and functioning.

The Homo Politicus often encounters laws, regulations, measures, and decisions that hinder the realisation of his goals. In the field of sustainability policy, it is quite common for existing law to enable or even promote climate-damaging technologies or excessive resource consumption. Therefore, Homo Politicus is interested in the constitutional state in the sense that she seeks to change or abolish certain laws and regulations so that better ones can replace them.

However, Homo Politicus understands that she can only act in a state structured by laws that regulate the coexistence of people. If such laws are lacking or have no validity, then the possibility of publicly advocating for justice is also lacking. In a state of lawlessness, state arbitrariness and private crime dominate, and actual political action becomes impossible. Therefore, it is essential for Homo Politicus to live in a constitutional state that establishes and protects certain freedoms for all citizens. For her activities, Homo Politicus needs a public forum where different and contradictory voices can peacefully argue about the res publica, the public matter.

Key Point: The Importance of Freedom, Equality and Rule of Law
A Homo Politicus must agree with his opponents that they are fighting as free and equal individuals for a meaningful cause; he must share with them an interest in ensuring that they and their successors can continue to fight, otherwise the existence of Homo Politicus himself is at stake. Therefore, Homo Politicus needs a legally ordered state, he needs a constitution that protects the freedom of conviction and provides rules of the game for all disputes. Many people in liberal constitutional states conceive constitutional democratic constitutions with their laws as a secure asset that will last for all times. The Homo Politicus, who is concerned with sustainability, must therefore also commit himself to the preservation of the legal state—as well as to the goals of major ecological transformations.

5.5 The Homo Politicus and the Right to Resist—Three Movements in the Fight for Climate Protection

Must Homo Politicus always adhere to existing laws? Three movements dedicated to the fight against climate change answer this question in different ways.

The movement *Fridays for Future* advocates for global adherence to the Paris Climate Agreement. However, the goals of the agreement are often not met, particularly in Germany, due to their association with significant economic cuts in sectors such as energy, industry, transport, construction, and agriculture. Fridays for Future justifies its demands with the urgent and crisis-like effects of climate change, such as droughts and floods, and calls for an ecologically moderate lifestyle. As the industrialized countries of the Global North are the largest contributors to anthropogenic, or human-made, climate change, while the developing countries of the Global South are most affected by its impacts, the movement criticizes the global state of "climate injustice". Fridays for Future operates within the framework of existing laws, using information campaigns and public protest actions.

The movement *Extinction Rebellion* is known for its spectacular acts of civil disobedience, such as traffic blockades and collective *Die-Ins*, where demonstrators lie down in public as if dead. In 2019, members of Extinction Rebellion chained themselves to the fence of the Chancellery in Berlin using bicycle locks around their necks. The Rebellion Day declared by Extinction Rebellion in 2018 caused a significant stir. Traffic on a Thames bridge was blocked by 6,000 demonstrators, and 85 people were arrested for traffic obstruction. By deliberately violating existing legal regulations, Extinction Rebellion aims to draw attention to the political and societal responsibility for the mass extinction of animals and plants and climate change.

The movement *Last Generation* poses the question on its homepage: "What will you do against the climate catastrophe?" This question is not directed at engineers who invent climate-neutral technologies, ministry officials who meticulously review administrative regulations for their suitability in the face of climate change, or consumers who strive to practice a sustainable lifestyle. Instead, it is directed at individuals who believe that appropriate action against the climate catastrophe involves "peaceful civil resistance against the destruction of our livelihoods". This resistance is justified with the statement: "For the necessary change, we must interrupt our entire daily routine and the entrenched fossil structures of our everyday life. Out of our love for all fellow human beings and out of our duty and responsibility for the well-being of the present and future generations." It should be noted that "risking a longer prison sentence" is listed on the homepage as a possible consequence of this resistance. This suggests that "peaceful civil resistance" systematically employs violations of existing law as a means to generate societal attention for its goals.[3]

Important to Know: Homines Politici in Civil Society Movements
Fridays for Future, Extinction Rebellion, and Last Generation all refer to scientifically verified findings regarding anthropogenic global warming. They concur in demanding societal awareness and a consistent policy that wards off the existential threat to humanity. All three movements act in the name of justice, demonstrate courage, engage with the public, and expose themselves to its criticism. Traits of Homo Politicus can be recognized in all three movements. However, they differ in their views on whether and to what extent violations of existing law serve the goals of sustainability policy.

From the perspective of Homo Politicus, it is crucial to consider that the entirety of laws, enacted through democratic procedures, constitute a legal state. This legal state is to be regarded as a highly valuable common good. If public awareness of the value of this common good diminishes, and if many members of society believe they can disregard law and order in favor of their own concept of justice, there is a risk that "might makes right". Imagine if not only climate activists, but increasingly other societal groups, would put legal breaches on their agenda for the sake of their goals. It could be expected that, in the long term, the "law" of the stronger would prevail. However, this is not law, but an expression of dominance, and ultimately, a state of general lawlessness could emerge. Unless such a state directly leads to the chaos of the "war of all against all", which philosopher Thomas Hobbes (1588–1679) viewed as the greatest evil in a political community, the rulers, powerful, and rich would likely benefit most from such a state, while the voice of the dissenters and weaker, unprotected by any law, would be silenced.

On the other hand, the existing laws of a state in their entirety almost never meet the requirements of justice, however one may define it. This is because they

[3] https://letztegeneration.de/mitmachen/, accessed on 05.05.2023.

express power and majority relations that were valid at the time they were decided. The approval they may once have carried belongs to the past. Thus, Homo Politicus sees a sometimes massive need for change. This is especially true today for a multitude of still valid regulations that allow or possibly even promote excessive resource and environmental consumption.

What can individuals in the role of Homo Politicus do in such a situation? Could targeted violations of existing law result in increased societal awareness of impending problems, thereby accelerating necessary transformation processes? However, every Homo politicus who systematically includes breaches of law in their actions must assume that their actions could serve as a role model for people with opposing views who identify themselves as Homines politici. These individuals might feel encouraged to break the law themselves. Moreover, it is somewhat contradictory to expect the state's organs, which are responsible for upholding the very law one disrespects, to support such actions. Finally, the question of power arises: Is it realistic to assume that calculated breaches of law will increase societal approval for a comprehensive sustainability policy?

From these considerations, it does not follow that resistance, as practiced by Extinction Rebellion and the Last Generation, should always be rejected from the perspective of the Homo Politicus. However, it does suggest that this resistance requires a stronger and more consistent justification than simple obedience to the law. The latter, even if practiced thoughtlessly out of sheer inertia, is unlikely to endanger the legal status quo as a whole.

Important to Know: Conditions for a Right to Resistance

A claimed right to resistance can only be convincingly justified if those who claim it can plausibly demonstrate that by breaking the law

- they are pursuing goals that are essential for the basic human rights of individuals or for the life of society as a whole,
- the legal status quo as a whole is not endangered,
- a societal consensus on the impending changes is accelerated,
- society can achieve the goals formulated in the name of justice much faster and more effectively than without the invocation of the right to resistance, and
- they are willing to accept the legal consequences of their law-breaking.

5.6 Summary and Conclusion

This chapter deals with concepts of man. The concept of man of conventional economics, Homo Economicus, was introduced and critically evaluated. With the assumption that people act as Homines Economici, individual behaviour in the economy can be influenced towards sustainable consumption—for example through financial incentives or monetary burdens. However, the concept is not sufficient when it comes to public affairs that concern the long-term common good. Here, the political dimension of being human comes into play, which is conceived with the Homo Politicus.

In contrast to Homo Economicus, Homo Politicus is interested in the common good and justice, and therefore also in the sustainability of a society's economic system. The empirical relevance of Homo politicus was shown, for example, in various studies on the behaviour of members of the ministerial administration and environmental agencies (Petersen et al., 2000). It was shown that private ambitions and interests, although certainly present, were not the only motives of the respondents.

Both Homo Economicus and Homo Politicus gain their plausibility from the fact that they correspond to our everyday experience. On the one hand, it is a basic fact that people follow their self-interest and seek their own advantage in a rational way. At the same time, however, we observe that people as Homines politici commit themselves in the long term to other people and creatures and to the common good. Although people as Homines politici can also act in a way that is detrimental to the common good, for example due to a lack of judgement, the dimension of Homo Politicus is crucial for a comprehensive sustainability policy. "The Homo Politicus as a representative of the pursuit of justice stands for the hope that a determined attitude on a path chosen with prudence and passion, with many small steps, could make life on earth, as far as it depends on human action, bearable for us and for future generations" (Manstetten, 2014, p. 31).

Summary: The Importance of Concepts of Man for Successful Sustainability Policy

In the conception of a practical sustainability policy, concepts of man play a crucial role. Homo Economicus and Homo Politicus refer to different aspects of being human. A person can embody both aspects in different fields of action. To understand processes in economy and society from the perspective of sustainability, one must draw on both the concepts of man of the Homo Economicus and that of the Homo Politicus. The Homo Economicus can be a helpful analytical basis when it comes to accurately assessing people's behaviour, for example as a reaction to the introduction of taxes and fees for individual environmental impacts. In addition, there must be people who, as Homines politici, commit themselves to public affairs and justice. This commitment is indispensable for the development and realisation of successful, long-term solution strategies (e.g., climate policy, energy transition) towards sustainability. Also in the public dispute about the best strategy, Homines politici are needed, who are concerned with the long-term approval and participation of the stakeholders, which make possible paths to a societal solution.

References[4]

Arendt, H. (1981). *Vita activa oder vom tätigen Leben [1958]*. Pieper.

Becker, G. S., Pies, I. & Streissler, M. (1996). *Familie. Gesellschaft und Politik – die ökonomische Perspektive*. Mohr.

Buchanan, J. M., & Tullock, G. (1962). *The Calculus of consent*. Logical foundations of constitutional democracy: The University of Michigan Press.

Krohn, P. (2023). *Ökoliberal: Warum Nachhaltigkeit die Freiheit braucht*. Frankfurter Allgemeine Buch.

Manstetten, R. (2000). *Das Menschenbild der Ökonomie: Der homo oeconomicus und die Anthropologie bei Adam Smith*. Karl Alber.

Manstetten, R. (2014). Nachhaltige Entwicklung in einer Welt von Dämonen? Max Webers Bild des Politikers und der Homo politicus. *Helmholtz Centre for Environmental Research, Discussion Papers 10/2014, Department of Economics*.

Mill, J.S. (1974). A System of Logic, Ratiocinative and Inductive, in: Collected Works, Vol. VII and VIII. University of TorontoPress.

Müller, D. (1989). *Public choice II*. Cambridge University Press: A revised Edition of Public Choice.

Petersen, T., Faber, M., & Schiller, J. (2000). Umweltpolitik in einer evolutionären Wirtschaft und die Bedeutung des Menschenbildes. In K. Bizer, B. Linscheidt, & A. Truger (eds.), *Staatshandeln im Umweltschutz. Perspektiven einer institutionellen Umweltökonomie*. Duncker & Humblot.

[4]The contents of this concept are based on: Faber, M., Frick, M., Zahrnt, D. (2019) MINE Website, Homo Politicus, www.nature-economy.com.

6 Responsibility: Who is Accountable and What Does it Mean When We take Responsibility for Environmental Problems?

Contents

What's it About?

This chapter is about responsibility, which is essential for solving environmental problems. It introduces a concept of responsibility that views responsibility as posing necessary questions to penetrate the complexity of environmental problems and develop practical approaches to their solutions.

The three central questions are:

1. Who is responsible for causing environmental problems?
2. Who is responsible for taking appropriate measures to solve and prevent environmental problems?
3. How can this responsibility be justified and its assumption demanded?

M. Faber et al., *Sustainable Action in Economy and Society*, https://doi.org/10.1007/978-3-662-69122-9_6

6.1 Introduction to the Concept

Until the 1970s, the influence of humans on nature was not a central theme in ethics. In classical ethics, nature is usually considered as something that lies outside the sphere of human action. Accordingly, the interactions between humans and nature not explicitly addressed. It was Hans Jonas who, in his groundbreaking book *The Imperative of Responsibility. In Search of an Ethics for the Technological Age* (published in German 1979 and in English in 1984), developed the foundations of an environmental ethics that deals with the fact that human behaviour has effects on the environment. Due to these effects of our actions, the environment falls within the realm of human responsibility. Rules for human action must be developed that ethically reflect and judge this action with regard to its effects on nature (see Faber & Frick, 2023). To perform this reflection and judgement, we must first understand to whom a particular environmental problem can be causally attributed, i.e., who has caused this problem through their action. This causal attribution seems simple at first, but is often difficult and forms the basis for the question of what moral and legal responsibility the actors bear for what they have caused with their action. Can they, for example, be morally or legally condemned for it?

Environmental problems are often so urgent, and the conditions and dangers associated with them so unbearable, that people must be found who take (or must take) responsibility for solving the respective problem. The first individuals to be addressed for taking on this responsibility seem to be those who have caused the problem through their action. However, since environmental problems often develop over very long periods of time, the responsibility for their elimination often has to be determined independently of those who caused it. A direct causal attribution is often prevented by the fact that the causers can no longer be clearly identified when the damage occurs much later. Some of the causally responsible people may even no longer be alive. In addition, it can be argued in their defence that they often could not have known at the moment of action what consequences this action would bring. We are often ignorant about parts of the consequences of our action (see Chap. 12, Ignorance). For example, at the beginning of industrialization, it was not recognized that the discharge of industrial wastewater into rivers would lead to a long-term die-off of the fish stock in these rivers. And when these damages were recognized, it was no longer traceable who had discharged wastewater when and in what quantities. A causal attribution of the damage to a causer could therefore no longer be made. The urgency and relevance of such considerations can be vividly seen in the current debate surrounding the climate issue. The purely causal attribution of responsibility often falls short when it comes to solving environmental problems. Therefore, further dimensions of responsibility must be developed and described, which is necessary to apply them meaningfully. We aim to do this step by step in the following sections.

6.2 Prerequisites for and Dimensions of Responsibility

Responsibility is a theoretical concept that plays a role in the fields of political philosophy, ethics, and law. As a term, responsibility is ubiquitous; it appears everywhere in practical life: someone takes responsibility or is under the demand to take responsibility. This is especially true in the environmental field, whether it is about wastewater, waste, or emissions. For this reason, an entire area of ethics of responsibility has developed. Hans Jonas laid the foundation for this with his aforementioned book *The Imperative of Responsibility* (1979 and 1984). With his "Imperative of Responsibility," he showed how people in general, and especially in the economy, and in politics, should behave ethically in relation to the environment.

The concept of responsibility refers to three principal properties of human action:

Important to Know: Prerequisites for Assuming Responsibility

- **Firstly:** The freedom of human action.
- **Secondly:** The ability of humans to realize their desires and intentions.
- **Thirdly:** The human ability to assess the consequences of one's own actions or to acquire knowledge about possible consequences of actions.

It should be emphasized that human influence on the environment, and our power over it through the development of modern technology in the last two centuries, has grown very quickly and dramatically. As a result, environmental destruction has increased significantly, especially from about the second half of the 20th century. Under these circumstances, how can the natural basis of human existence be preserved? This is the central question of sustainability (see Chap. 4 *Sustainability and Justice*), for which the concept of responsibility is necessary to provide an answer.

Free persons can choose their goals autonomously and realize them within their possibilities. The corresponding actions and their consequences can be *attributed* to them, and in this sense, they are responsible for their actions and their consequences. Consequently, we can initially make a *causal attribution* of an action and its consequences to a person. For example, a minister can say, "I take responsibility for the consequences of a measure I ordered." We have often heard this phrase during the times of the Covid-pandemic. However, we have also repeatedly experienced that this first step does not necessarily lead to consequences for the *causers*; because mere attribution does not mean, for example, that the minister has to resign or personally pay for the elimination of a possibly caused damage.

For our question about responsibility, this first step of attribution is absolutely necessary, but by no means sufficient. Beyond attribution, we need to assume responsibility in the sense of a *legal and moral obligation*. We recognize that this legal and moral obligation can only occur if the causal attribution has been determined beforehand and if we are dealing with free people. If persons were forced to act, the consequences of this action can indeed be *causally attributed* to them. However, a moral or legal responsibility cannot result from this, as it crucially depends on the person being able to freely decide for or against an action. We recognize that this second, legal and moral obligation has much stronger implications than the first, which initially only describes relationships and does not have to lead to consequences on its own. We add that for legal and moral responsibility, it cannot be thought of separately from other people, the community, or society. Responsibility requires a counterpart or an authority that names and demands legal or moral responsibility. For example, as members of a society, we are responsible to this society for complying with the applicable legal rules, an employee is responsible to their superiors for fulfilling their tasks, the government is accountable to the parliament.

At the same time, there are also limits to individual responsibility. When we think of natural disasters such as the flood disaster in the German Ahr Valley in 2021, it becomes apparent that there are situations in which the responsibility of the individual is limited by ignorance (see Chap. 12) or uncontrollable circumstances such as extreme weather, an earthquake, or the unfortunate combination of circumstances, so that the state must assume responsibility.

Important to Know: Limits of Individual Responsibility

Individuals' assumption of responsibility reaches its limits where

- People cannot freely decide on their actions.
- People are confronted with ignorance about the consequences of their actions, which they cannot reduce through their own efforts.

This necessity requires politicians to assume *political responsibility* that goes beyond the limited scope of individual responsibility constrained by ignorance and lack of control. But how can a politician deal with this ignorance or the unfortunate combination of circumstances? Firstly, politicians have different access to knowledge and support in finding solutions to complex problems and relationships. They can consult scientists, commission expert opinions, or finance research projects for specific questions. Nevertheless, it is often not easy to determine or even predict which consequences and damages certain actions will cause, which decision should therefore be made with regard to the political area of responsibility.

In addressing this question, two extensive philosophical fields emerge: the previously formulated challenge of *ignorance* and the category of *power of judgement*. We aim to focus on the fact that dealing with ignorance is a central task for politics. One method of addressing this challenge is the application of *power of judgement*. What does this mean? Power of judgement allows for a *good decision* to be made in the interest of the matter at hand, despite prevailing ignorance and therefore on the basis of incomplete information. However, the difficulty lies in the fact that power of judgement cannot be taught, but can only be gained through experience (see Chap. 7 *Power of Judgement*).

Important to know: Three dimensions of responsibility

When discussing responsibility, it is important to distinguish between different dimensions:

- **Causal Responsibility:** This describes the connection between specific individual actions and the consequences that are *caused* by these actions. This is initially an *attribution*, which does not yet involve a moral or legal *judgement*.
- **Moral and Legal Responsibility:** If actions can be causally linked to certain consequences, the decision to take these actions and the corresponding consequences can be *evaluated* and *judged* using moral or legal standards. *Moral or legal responsibility* can then be attributed to the causer. This judgement can have consequences.
- **Political Responsibility:** Individual responsibility reaches its limits where people cannot freely decide on their actions or encounter irreducible ignorance regarding the consequences of their actions. When individuals cannot be attributed responsibility due to these limits, political responsibility comes into play. Political actors, by virtue of their office, have greater influence and better access to knowledge, which is why their scope of responsibility is defined more broadly. In order to make good decisions despite high complexity and ignorance, politicians need power of judgement to assume political responsibility.

6.3 The Interplay of Different Levels of Responsibility

How does the concept of *responsibility* help us to understand concrete environmental problems, and to develop solutions? We aim to answer this question using the example of the pollution and subsequent renaturation of the Emscher river over more than 200 years.[1] The historical contexts and processes are structured using the concept of *responsibility*, which here appears in close interaction with the concepts of *time* (Chap. 8), *joint production* (Chap. 13), *ignorance* (Chap. 12) and *power of judgement* (Chap. 7).

Firstly, it is necessary to understand why the Emscher river is a meaningful example: The Emscher, from its source near Dortmund to its mouth in the Rhine near Dinslaken, passes through large parts of the Ruhr area. On its journey, it passes cities such as Gelsenkirchen, Bottrop, Oberhausen, and Duisburg—all known for mining as well as iron and steel production. With its approximately 100 km length, the Emscher flows through an area that is, so to speak, the cradle

[1] We first used the example of the Emscher in a discussion paper to illustrate complex relationships in environmental issues (Faber et al., 2021)

of German industrial history. In its catchment area of 865 square kilometers live about 2.3 million people—a conurbation, emerged and shaped in the context of an industrialized modernity.

The ecological problems evident in the case of the Emscher are typical of those that have occurred and continue to occur at heavy industry sites worldwide. In 1800, the Emscher was a largely untouched natural watercourse. However, it was subsequently increasingly used, polluted, and modified by agriculture and, ultimately, industrialization. It gained notoriety as "Germany's dirtiest river" (FAZ 2015)[2] and was eventually cleaned and restored in an expensive and elaborate process spanning more than two decades until 2020.

6.3.1 Characteristics of Environmental Problems: Long Timeframes and Joint Production

To understand the interrelationships, two concepts from MINE should be mentioned first, the consideration of which is crucial for a structured approach to the problem and the subsequent assignment of *responsibility*: The concepts of *time* and *joint production.*

The pollution of the Emscher is caused by joint products of human coexistence and economic activity, which interact with each other over a very long *period of time* and the removal of which also takes a very long time. With the development of the Ruhr area as a mining and industrial region from 1850 onwards, not only were large production facilities built, underground coal mines established, and transport routes opened up, but hundreds of thousands of workers and their families also had to be settled and supplied with the means of daily life. Consequently, the region's rivers mixed industrial waste and mining residues with human waste and faeces. All unwanted joint products of human life and activity, were, unlike the intended products such as iron and steel, not in the focus of anyone's attention (see Chap. 13 *Joint Production*). However, they destroyed the fauna and flora in the rivers and caused infectious diseases such as typhus and cholera to spread in the adjacent settlements during floods.

6.3.2 Ignorance as a Limit to Responsibility

The described conditions are consequences of human actions, which resulted over a longer period of time from the interaction of different joint products and which were given little attention for a long time or about which ignorance prevailed (see Chap. 12) until they became recognizable in the form of negative impact for

---

[2] Andreas Rossmann in the article "Renaturierung der Emscher. Das ist Köttelbecke" from 30.07.2015 (https://www.faz.net/aktuell/feuilleton/deutschlands-schmutzigster-fluss-emscher-wird-renaturiert-13725603.html). Last accessed on 18.06.2023.

people. This ignorance about the complex and dangerous interplay of joint products leads to the question of responsibility for the ecological consequences and their removal being asked concretely only much later.

In 1899, almost 50 years after the opening of the first mines, not only the Emscher but also the other rivers of the Ruhr area were essentially open sewers. And even though the citizens, companies, cities, and municipalities suffered from the condition, for many years no one took responsibility for improving the situation.

This is not particularly surprising, considering the conditions for taking responsibility:

- First, the causal relationships must be correctly assigned, i.e., it must be clarified whose action is causative for specific consequences.
- If this is successful, the question must be asked whether the causers can actually be held morally or legally responsible, i.e., obliged to take responsibility for their actions and the corresponding consequences.

That these conditions can be met is rather unusual in the case of complex environmental problems. Moreover, the causal attribution is already challenging due to the long periods of time it takes for environmental problems to become recognizable as such, and because of the interaction of various joint products with each other. In addition, for the assumption of moral and legal responsibility, it is important to consider that people can only be held responsible for those consequences of their actions of which they can know or with which they must reckon.

If there is complete ignorance about specific consequences and effects of actions in advance, the attribution of individual responsibility reaches its limits. The example of the Emscher shows how prevalent ignorance was in the understanding and handling of changes in the environment in the 19th century. Therefore, hoping for individuals or individual companies to take responsibility is neither promising from pragmatic nor from responsibility-theoretical points of view.

6.3.3 Institutions as Bearers of Political Responsibility

In 1899, the so-called Emscher Cooperative was founded, which the affected cities and municipalities were required to join. The Emscher Cooperative represents an entity that, beyond private and economic interests, assumes responsibility for addressing a problem that is both socially and ecologically urgent. Not individual people or the industrial companies responsible for the issue intervene, but a higher-level *political institution* assumes the *political responsibility*.

The Emscher Cooperative was tasked with developing a concept for wastewater management. The decision was made to transform the Emscher from an independent river course into the central drainage canal of the Ruhr area. The rivers Ruhr and Lippe were used for fresh water supply. The Emscher was canalized and shortened from its original 109 kilometers to a river course of 83 kilometers, and the flow speed was significantly increased to quickly drain the wastewater, which could not be led underground due to the many active mines.

Technically, this strategy worked well throughout the 20th century, but it was an ecological and hygienic disaster. A stinking, highly toxic cesspool ran through the Ruhr area. The animal and plant life on the river died, and for the people, in addition to the odor nuisance, there were significant health risks. The high concentration of heavy metals can cause complaints such as sleep disorders, respiratory diseases, skin diseases, intestinal diseases, and the like in people who come into contact with the water. In adolescents, heavy metals such as lead or chromium can damage the circulatory system and the central nervous system, leading to severe learning difficulties.

These health risks led, in 1991—almost 200 years after the beginning of human interventions in the Emscher river system and almost 92 years after the foundation of the Emscher Cooperative—to the initiation of cleaning and restoration efforts for the Emscher. The starting point was the *European Water Framework Directive* of the European Union, a binding framework for national legislations, which gradually introduced strict limit values for the pollution of waters. The condition of the Emscher, had it remained unchanged, would have been evaluated by Brussels as a blatant violation of the limit values of the Water Framework Directive by 2015 at the latest and possibly punished with high fines. The assumption of political responsibility on site prevented this.

6.3.4 Power of Judgement—How Responsible Action Can Succeed Despite Ignorance

Since environmental policy decisions, due to insurmountable ignorance, are almost never made on the basis of a complete factual basis, a decision for the "perfect solution" is usually not possible. This is where the concept of *power of judgement* (see Chap. 7) comes into play, the practical ability to respond appropriately to the respective situation-specific challenges and to remain capable of making decisions under the given circumstances. Power of judgement means *doing the best under given circumstances with given knowledge.*

The actors of the Emscher Cooperative have shown power of judgement, even if the decision to turn a living river into a dead sewage canal may seem shocking from today's perspective. But considering the knowledge available in 1900 and the interests and power constellations at that time, it is not entirely implausible that they made the best possible decision under the given circumstances.

Power of judgement can be particularly attributed to the authors of the European Water Framework Directive, who explicitly placed ecologically sensible river basin management at the center of their objectives and provided this ambitious objective with the necessary long time horizons. For the necessary changes on site, more than 20 years were given already in 1991 by the early announcement of the target date 2015. Taking the appropriate measures remained a challenging task, but also a task that could at least be solved for the Emscher.

In 2011, 211 years after the first interventions in its river course and 111 years after the Emscher was officially declared the sewage canal of the Ruhr area, a fish was sighted for the first time in many years in its upper course near Dortmund. Today, a popular cycle path runs along the banks of the Emscher that combines leisure in the green with a journey through German industrial history.

References[3]

Faber, M., & Frick, M. (2023). Environmental Ethics. In B. M. Haddat & B. D. Solomon (eds.), *Dictionary of ecological economics. Terms for the new millennium* (S. 197–198). Elgar.

Faber, M., Frick, M., & Manstetten, R. (2021). Die Online-Plattform MINE-eine Brücke zwischen Umwelt und Wirtschaft. *AWI Discussion Paper Series, 701.*

Jonas, H. (1979). *Das Prinzip Verantwortung-Versuch einer Ethik für die technologische Zivilisation.* Suhrkamp. [English Translation published in 1984: The Imperative of Responsibility. In Search of an Ethics for the Technological Age. Chicago Press]

[3] The contents of this concept are based on: Faber, M., Frick, M., Zahrnt, D. (2019) MINE Website, Responsibility, www.nature-economy.com.

Power of Judgement: How to Proceed when we Reach the Limits of Our Knowledge? From Knowledge to Action

7

Contents

What's it About?

This is about knowledge and the question of what we can know. To achieve sustainability goals, a society needs not only a firm will and reliable knowledge but also the ability to translate knowledge into appropriate action.

Scientific knowledge, even if it provides an accurate description and analysis of the facts, is not sufficient for effective action towards sustainability. Effective action requires *practical knowledge,* the know-how of actors in politics, economics, and society. The connection of scientific knowledge with practical knowledge requires a competence that is introduced in this chapter with the term *power of judgement.*

M. Faber et al., *Sustainable Action in Economy and Society*,
https://doi.org/10.1007/978-3-662-69122-9_7

7.1 Introduction to the Concept

Comprehensive solutions in areas such as climate, waste prevention, protection of biodiversity, or use of natural resources could, it seems, be rapidly and decisively advanced if we only followed the recommendations of science. For the climate problem, the following solution seemingly is a good choice: Natural scientists provide reliable information about the sources of greenhouse gases and their effects in the atmosphere, engineers design technologies for climate-friendly production processes. Economists, social scientists, and jurists develop proposals for commands and prohibitions or market-conforming levies and certificates, and politics implements what the experts from the sciences demand. Should not all this lead to climate neutrality being achieved in a relatively short time?

Upon closer inspection, however, a completely different impression emerges: Even if there is broad agreement on the problem description by the natural sciences, debates about solutions are controversial, the processes of decision-making in politics and society are protracted, and the results achieved prove to be contestable compromises between environmental protection and lobby interests. Pessimists therefore assume that humanity has the knowledge to preserve its livelihoods, but fails to translate knowledge into action.

But this impression is not accurate, at least not always. The environmental policy of the last fifty years in Germany has indeed achieved significant changes, especially in the areas of wastewater and waste, the phase-out of nuclear power has been initiated—even if it took a long time—and the impending end of lignite mining seems irreversible. Examples such as the worldwide ban on CFCs—which demonstrably damaged the ozone layer—or the implementation of the European Water Framework Directive show that effective solutions can also be found for complex, international problems. So, environmental problems can indeed be tackled with a prospect of success. Although the path from the first problem identification to a political solution is often arduous, and the distance is usually much longer than initially thought, it has been shown in principle that successful environmental political action is possible.

The following is about the transition between knowledge and action. At the interface stands the ability of *power of judgement*. This mediates between the information and models of the theoretical sciences and the special requirements that arise from the action situation, its political, economic, and social circumstances.

Important to Know: The Difference between Know-that and Know-how

Scientific knowledge contains information about reality that is integrated into models and theories. This knowledge is a *Know-that*. When experts from the various sciences are consulted by decision-makers as advisors on certain problems, the actors cannot directly implement the Know-that of the specialist sciences into reality. For implementation and possible adaptation, they need a special *Know-how,* a knowledge, *how to do it,* so that the right theories in the respective political, economic, and social circumstances lead to good results in practice.

7.2 Power of Judgement as Know-how[1]

The Know-that of the individual scientific disciplines can be presented in a transparent way in concepts and coherent arguments. Knowledge that is suitable for practice has a different character. A successful football coach must (in our time) have a lot of theoretical knowledge, but for the lineup of her team and for her tactical arrangements with regard to the upcoming game, she needs skills that she cannot learn from any sports science textbook: The quality of a coach is measured by her practical knowledge, her *Know-how* (Klauer et al., 2017, p. 101).

Power of judgement is expertise, practical knowledge, in conjunction with the theoretical knowledge required for a decision-making situation.[2] But expertise is only in rare cases simply the schematic application of theoretical knowledge. In complex decision-making situations, even the most thorough theoretical analysis in advance cannot reliably indicate which option will lead to the desired result. What a *correct* choice is, cannot be said with scientific accuracy before the occurrence and analysis of the consequences of action. People who manage to choose a goal-oriented course of action in such decision-making situations can be attributed *power of judgement.* They make a choice that, in retrospect, becomes recognizable as an *appropriate* choice, even if there were reasons for other options beforehand. Whoever has power of judgement, in addition to the necessary theoretical knowledge, has a *lucky hand.* But although the success of actions guided by power of judgement indeed also depends on the actors not being downright unlucky, the luck needed for success is the luck of the capable. *Power of judgement is ability.*

7.2.1 What Power of Judgement Achieves—Immanuel Kant's Example of Medical Competence

What power of judgement is and what it achieves, Kant makes clear with examples: "A doctor, therefore, a judge, or a statesman can have many beautiful pathological, legal or political rules in his head, and yet he will easily violate them

[1] The following considerations are owed to Klauer, Manstetten, Petersen and Schiller (2017, Chap. 7).

[2] The following remarks refer to Kant's doctrine of power of judgement. However, the concept of power of judgement in Kant's work has a multitude of implications that cannot be taken into account in our highly simplified presentation. See Wieland (1998) for this.

in their application, either because he lacks natural power of judgement, and he can see the general *in abstracto*, but cannot distinguish whether a case *in concreto* belongs to it, or also because he has not been trained enough by examples and real affairs for this power of judgement" (Kant, 1977, p. 173).

The example of medical competence shows that power of judgement depends on two characteristics:

1. Indispensable is the knowledge of medical expertise, the know-that of conventional medicine as well as serious alternative offers. Such knowledge is conveyed in textbooks, lectures, and professional conferences.
2. Equally important is the ability to develop an eye for the respective patients, their overall physical condition, their life situation, their family environment, their professional stress, etc.

A doctor must therefore connect the particular case with her theoretical knowledge. In some cases, she will judge that the concrete symptoms of the examined individual correspond to the general description of a certain disease pattern. Then she can prescribe a conventional therapy. Sometimes, however, her eye for the patient leads her to decide on a treatment that seems to contradict what is written in the books. This does not speak against the books, in which rules of diagnosis and therapy can only be formulated generally and abstractly. However, in each concrete case and its specific circumstances, additional aspects come into play—sometimes more, sometimes less relevant. The consideration and appropriate weighting of these special aspects in combination with a scientific know-that at the level of available factual knowledge: This is what constitutes power of judgement in the field of medicine.

Starting from this example, power of judgement can be described as an eye for the requirements of the respective special circumstances, which includes all available scientific knowledge, but does not stop there. Accordingly, Kant formulates: "Power of judgement [is] the ability to subsume under rules, i.e., to distinguish whether something falls under a given rule or not" (Kant, 1983 [1781] p. 184, quoted from Klauer et al., 2017, p. 98).

By *rules*, Kant understands all abstract knowledge and all ethical orientations that could be relevant for actors in a certain case. But then the question arises as to which of these rules need to be considered in a specific situation and which additional aspects of this situation are not covered by any rule. There "exists a gap between our knowledge and its application in individual cases, which cannot be closed even by an ever-increasing specification of this knowledge. Application and interpretation of scientific knowledge requires a kind of knowledge that we can call *practical knowledge*" (Klauer et al., 2017, p. 115).

Important to know: When power of judgement is needed

Power of judgement is needed whenever we have to decide whether and to what extent our scientific knowledge (know-that) can be applied to a specific case in practice. And we still need judgement to decide what this practical application should look like in concrete terms (know-how).

7.2.2 Ignorance, Complexity and Capability to Act

Power of judgement in the fields of environmental and sustainability policy requires special skills to handling the relevant theoretical knowledge, as this knowledge is often incomplete. A significant portion of such ignorance in environmental problems arises from the fact that the interaction of natural and human systems must be included in the problem description, which is never entirely possible. This is particularly true when considering medium and long-term interactions between processes in politics, economy, and society on one hand, and natural processes in ecosystems on the other. Efforts to gain comprehensive and consistent knowledge quickly reach their limits (see Chap. 12 *Ignorance* and in Chap. 13 *Joint production*).

We are dealing with a wide range of different knowledge areas and forms of knowledge: Chemical, biological, ecological, and physical insights are as relevant as knowledge about behavioural psychology, legal mechanisms, economic incentive systems, and political science knowledge about democratic processes for the adoption of certain rules. The various scientific findings cannot usually be combined into a whole that would serve as the scientific basis for action options. This applies not only to the *diagnosis* of environmental problems in their natural and social dimensions, but also to *therapy*, the elimination and future avoidance of an environmental problem. This introduces another task to the questions of scientific knowledge: The question of what appropriate actions can look like and the question of how such actions can be initiated. If the patient is the whole society, the term therapy encompasses a complex of political, legal, economic, and social contexts that need to be considered. If the treatment of a single person occasionally poses great problems for doctors, then it is infinitely more difficult to guide an entire society onto an environmentally "correct path". The complexity of environmental problems in a complex social world makes it impossible to make decisions with scientific accuracy. Too many areas of knowledge and systemic interrelationships need to be taken into account for clear cause-effect relationships to be formulated between the various concrete action options and the respective results that would result from them.

In such decision-making situations, power of judgement is required. However, what usually characterizes power of judgement is not sufficient here: the meaningful connection of theoretical knowledge with the requirements of practice. This is because the status and the range of theoretical knowledge must also be considered. Here, power of judgement must prove itself in dealing with ignorance. In order to remain capable of action, the actors need courage and confidence in advance to be able to act concretely into the unknown despite all ignorance, and afterwards self-criticism, i.e., the willingness to recognize and admit mistakes and to correct wrong decisions.

Important to Know: Ignorance and the Impossibility of a *Perfect Solution*
Since environmental policy decisions can rarely be made on a complete factual basis due to insurmountable ignorance, a decision for the "perfect solution" is usually not possible. This is where *power of judgement* comes into play, the practical ability to respond appropriately to the challenges specific to each situation and to remain capable of making decisions under the given circumstances, the context. Power of judgement means doing what is appropriate and suitable—given the circumstances with the given knowledge.

7.3 Power of Judgement and Transformation—The Need for a Comprehensive View and Recognizing the Right Time

However, the power of judgement on the part of the respective actors is not sufficient when it comes to comprehensive and profound transformations. A whole society, in which these transformation processes take place, also needs judgement. A certain critical distance to day-to-day politics must, when it comes to comprehensive changes, be supported by a basic trust in the actions of the responsible actors. It can already be problematic if, for example due to media reporting, the opinion arises that the decision-makers do not have enough knowledge and experience for the upcoming tasks. It becomes downright dangerous when politicians are generally accused of acting solely on the basis of private self-interest or as agents of inscrutable powers. If such opinions become dominant, the ability of a society to act as a whole is put at risk.[3]

Power of judgement in matters of transformation involves, on one hand, the ability to comprehensively understand problems. On the other hand, judgement also includes a sense of timing, knowing when to act and when not to. A sense of time is of utmost importance here (see Chap. 8 *Time*). The ability to identify the *kairos*—the right moment, the opportune window, the time that is different because it makes possible what usually seems impossible—is a skill that cannot be universally taught through textbooks. Instead, it relies on experience, attentiveness, and intuition. In short, those who can recognize the right time and act accordingly demonstrate power of judgement. From the perspective of power of judgement, it is important to understand that the future can never be controlled. We discuss this in Chap. 12 under the term *ignorance*: The future is open. Regardless of our intentions and goals, we must acknowledge that we can never predict all outcomes—things often turn out differently than expected. Power of judgement, in this context, is the ability to remain open to novelty, enabling one to creatively handle surprises.

Power of judgement requires experience in reality and concrete examples. Transformations are not utopias that start from scratch. Rather, we are already in

[3] In the last section of this chapter, we will show, using the example of the Citizens' Councils Climate, how the formation of power of judgement can be enabled in a societal process.

the midst of a process of major transformations, with both successes and setbacks. Examples of both are provided, as we can learn from each.

7.3.1 An Example of Power of Judgement Facing Difficulties

Under increasing pressure from environmental associations, and more significantly due to the sharp rise in oil prices in the 1970s, the automotive industry has successfully pursued the development and use of more economical engines. From an environmental policy perspective, the prospect of significantly reducing fuel consumption of motor vehicles seemed desirable. However, in retrospect, it is clear that the substantial increase in combustion engine efficiency has not led to a decrease in the absolute amount of fuel required in the mobility sector. In advance, the knowledge (know-that) about the technological interrelationships suggested that the use of more economical cars would reduce society's overall oil consumption. However, looking back, we see that technological development has led to a change in people's behaviour—and not in a desirable way. The savings achieved resulted in more cars (keyword: second car) and also larger and more powerful cars being demanded. This trend continues to this day. For the maintenance of cars is cheaper than before, as they consume far less fuel. Through a so-called *rebound effect*, the increase in efficiency and fuel saving in each car has been overcompensated; it therefore did not lead to a reduction, but to an increase in total fuel consumption. Once observed and described, the insights about the complex interactions of technological developments and behavioural changes can be taken into account for power of judgement in decision-making. However, the knowledge of when these types of interactions occur or not remains, due to the complexity of the relationships, less precise than clearly formulated causal chains. This example shows that the application of power of judgement must be subject to a constant learning process.

7.3.2 An Example of Environmental Protection without Power of Judgement—The Clean Water Act of 1972 in the USA[4]

In 1972, the American Congress passed an ambitious water law, the *Clean Water Act* (hereafter abbreviated as CWA). The then burgeoning environmental movement had a significant influence on its drafting (Roberts, 1974, pp. 18–21). The goal outlined in the CWA was to prohibit any discharge of pollutants into the waters of the USA by 1985. A critic characterized the approach of this water law as follows: "The existing US system for controlling water pollution is dominated

[4]The explanations in this and the subsequent section are derived from the essay by Binswanger, Faber and Manstetten (1990).

in two respects by an approach that focuses solely on prohibitions and threats of punishment. Firstly, it emphasizes the total prohibition of wastewater discharges into public waters, instead of applying cost-benefit principles (...). Secondly, the US system relies heavily on the threat of penalties, i.e., fines and/or imprisonment, and not on economic incentives, to induce industries, municipalities, and other waste producers to reduce their pollutant discharges into public waters" (Brown & Johnson, 1984, p. 952). The CWA was unanimously passed in the US Senate. However, there was no corresponding consensus among the parties, associations, trade unions, administrations, and representatives of various sectors. Thus, the prerequisites for broad social acceptance were lacking. Such acceptance was hardly to be expected given the consequences that would have resulted from the realization of the ideal solution aimed for in the CWA. Because the implementation of the prohibition of any discharge of pollutants into the waters of the USA would have led to a near-crisis restructuring of the American economy. This would have resulted in, among other things, the bankruptcy of numerous sectors of the economy, particularly the chemical industry, mass unemployment, and impoverishment of broad sections of the population. It turned out that it was impossible to enforce the CWA even to a small extent.

Measured against the goals outlined in the CWA of 1972, water legislation in the USA in the seventies proved to be a failure (OECD, 1987, p. 54). The apparent complete success of the environmental movement was, in terms of the actual goal, a Pyrrhic victory: instead of zero water pollution, everything remained as it was. These were the consequences of an ideal solution, in the development and enforcement of which power of judgement played hardly any role.

7.3.3 An Example of Environmental Protection through Power of Judgement—the Water Legislation in Germany in 1976

A counterexample to the CWA is the water legislation in the Federal Republic of Germany in the mid-seventies, where the creation of a consensus was crucial for its relative success. The initial proposals for legislation to combat water pollution suggested that fees should be levied directly from the dischargers of waste into waters, adhering to the *polluter pays principle*[5]. However, some federal states, particularly Bavaria and Baden-Württemberg, rejected these proposals as too radical. Instead, they recommended a more moderate fee system, which would work in tandem with the traditional standards and regulation system. By 1976, the idea of a combined system of regulations and fees had prevailed. This system was

[5] The polluter pays principle is a fundamental concept in environmental policy, stating that those responsible for environmental pollution and degradation should bear the associated costs. In other words, those who cause environmental damage are responsible for the elimination or repair of this damage. This approach creates an incentive to reduce or completely stop environmentally harmful behaviour, as the polluters must take direct financial responsibility.

designed to levy fees that were high enough to incentivise the reduction of environmental pollution, while maintaining an administrative system to control it. The industry initially rejected the idea of a wastewater fee system. However, as political support for the system gained momentum, resistance shifted to implementation issues, such as the criteria for setting the fees, the amount of the fees, and the dates when the system should come into effect (Brown & Johnson, 1984, pp. 931–932). All such questions were pragmatically resolved. For instance, the fees were initially kept low and only gradually increased over the course of years. The increase was announced from the beginning, so that the affected companies, municipalities, and water associations could adjust to it in time. This approach ensured broad acceptance of the measures, which demonstrated the power of judgement of the decision-makers.

This type of law was successful not only in comparison to the USA, but also in absolute terms, as the Federal Republic of Germany had the highest percentage of biological or chemical treatment plants of all countries by the end of the eighties, alongside Denmark and Sweden (OECD, 1987, pp. 53–54). It was not so much the individual aspects of the law as the interaction of economy and politics that were important for the effectiveness of the environmental measures at that time. The relative success of German water legislation compared to American was the result of a consensus reached after years of controversies and discussions among many societal groups, such as science, industry, trade unions, associations, administration, and representatives from politics and parliaments. Such debates are necessary to identify problems, neglected aspects, and disadvantages, so that a common basis for action can be reached after such a clarification process. Only when a consensus is found among such groups can one hope for the acceptance of such important laws by the majority of the population.

7.3.4 Activating the Power of Judgement of Civil Society—the Citizens' Council on Climate

Important to Know: When Power of Judgement Can Become Effective

Power of judgement can only become effective in a society when

- The actors act in the sense of the *Homo Politicus* (see Chap. 5) and
- Society *trusts* those involved to act according to Homo Politicus.

Where both are given, political action complements societal acceptance, which it relies on.

Immanuel Kant and Hannah Arendt have highlighted the importance of a communal sense for the formation and exercise of power of judgement. This can be seen as a process: where power of judgement is exercised, the actors as Homines

Politici strive for approval by civil society. On the other hand, the inclusion of civil society in decision-making processes can ensure a certain social acceptance from the outset. We want to finally present a concrete attempt that aims to make the competences of citizens fruitful for dealing with climate change. It is the *Citizens' Council on Climate*.[6]

The Citizens' Council on Climate was initiated in December 2020 by the association *Citizens' Initiative for Climate Protection* (BBK) and the *Scientists for Future*. Former Federal President Horst Köhler, as patron of the Citizens' Council on Climate, stated its objective: "Can we manage to maintain the living conditions on our planet in the long term and in a way that is acceptable to all people by changing the way we produce, consume, live? If Germany wants to achieve the goals it committed to in the 2015 Paris Climate Agreement, a great societal willingness to change is needed. That's why it's so important that citizens are involved in the search for solutions—and that politics takes their proposals seriously." For the formation of the Citizens' Council, several thousand citizens were first contacted with randomly selected telephone numbers in a staggered lottery procedure. From this group, 160 participants were then selected in several steps and according to predefined demographic characteristics so that they represented German society as well as possible. The Citizens' Council was scientifically advised by a board of scientists from various disciplines as well as experts from the different subject areas of the Citizens' Council. This board of the Citizens' Council on Climate consisted of selected representatives from business, social and environmental associations, civil society, religious communities and other interest groups. They were to represent civil society, observe the process and ensure content balance.

In the spring of 2021, the content framework was developed in which the Citizens' Council should develop its recommendations. Proposals from scientists, politicians and organizations as well as a nationwide opinion poll served as the basis. The members of the Citizens' Council could also suggest topics during the consultations. They participated in twelve sessions between April 26 and June 23. Visions of the future and guiding principles were the preliminary result of the first three sessions. The recommendations from the working groups were finally voted on in the plenary and a final version of the visions of the future and guiding principles was formulated. The procedure and the results were compiled after the last session in the report *Our Recommendations for German Climate Policy*, which was handed over to all parties represented in the Bundestag in the fall of 2021.

After the implementation phase, participants of the Citizens' Council highlighted learning effects and the search for compromise. Valid information was important for their decisions. Earth system scientist Wolfgang Lucht pointed out

---

[6]The following remarks represent, with few modifications, the wording of the (greatly shortened) press release of the Citizens' Council on Climate: https://buergerrat-klima.de/presse/pressemitteilung-1 (accessed on 18.04.2023). We have refrained from specifically marking the abbreviations.

that the results of the Council do not reflect the views of an interest group, but are the result of a free negotiation process. Political scientist Ulrike Zeigermann and health scientist Stefanie Ettelt also examined how such publicly developed approaches can contribute to solving complex problems using the example of the Citizens' Council on Climate. The complex questions around climate change and climate policy were reduced to manageable levels in fields of action. Nevertheless, the participants were sometimes overwhelmed by the complexity. They concluded that the practical handling of climate change by citizens contributes to legitimate political solutions. However, such deliberative and resource-intensive initiatives remain relatively subordinate, with limited visibility and limited influence on politics.

7.4 Concluding Remarks

Discourses that affect society as a whole can lead astray for a variety of reasons. We wish to highlight two dangers. The first is the belief in science as a panacea for all societal ills. The second, and perhaps even greater danger, is hostility towards science, where the mere consensus of scientists on certain theses can arouse suspicion. Those who oppose science often rely on inspirations, intuitions, or mere gut feelings, if not directly deriving their knowledge from manipulative sources within the internet's filter bubbles. Power of judgement serves as the middle ground between these two extremes, helping to avoid their pitfalls.

While a degree of gut feeling is integral to power of judgement, the decisive factor is not the gut feeling itself, but the right blend of scientific knowledge, intuition, context, and willingness to self-correct. This is complemented by an openness to the perspectives of others who have valuable insights on the issues at hand. Given these prerequisites, one might question whether there is anyone who possesses sufficient power of judgement. This question can be answered affirmatively when we consider that power of judgement is not a static quality, but one that can be continually developed through long processes of *trial and error*. Power of judgement is not a one-time acquisition, but a quality that can be improved upon. Experience is the best guide in this regard. If we view power of judgement as more of a process than a result, it is not as rare as one might think. Anyone who acknowledges their ignorance yet remains interested in participating in upcoming changes requires power of judgement and understands its necessity. This awareness of the need for power of judgement forms a solid foundation for all individuals committed to the fields of environment and sustainability.

Given the current state of discussion, it can be concluded that much is already gained if all participants in political processes understand that the factor of power of judgement is indispensable.

Summary: Some Rules of Power of Judgement

- Strive for consensus (in the sense of broad societal acceptance).
- Develop and maintain self-confidence, inspire others with your confidence, and thereby broaden the basis of societal trust towards consensus.
- Always adapt principles to circumstances, but never lose sight of them (avoid both dogmatism and opportunism).
- Major transformations are processes that require both patience and tenacity.
- Consider appealing solutions, but never be seduced by them. After careful consideration, you may not know what is possible, but you can certainly identify what *does not* work. Recognizing and excluding dead ends are crucial elements of the decision-making process.
- Compromises, even painful ones, are necessary, but one must recognize when a compromise not only *delays* (delay is part of the political process and inevitable), but permanently *prevents* the achievement of a goal.
- Always remain self-critical—show interest in the criticism of others and take it seriously—and yet do not be irritated by it.
- The goal is not the optimal solution, but—at best—the good, and—at worst—the halfway tolerable solution. Achieving the latter can sometimes be a great success. However, when the halfway tolerable solution is presented as the optimal one, any transformation is compromised.

References[7]

Binswanger, H.-C., Faber, M. & Manstetten, R. (1990). The dilemma of modern man and nature: An exploration of the Faustian imperative. *Ecological Economics, 2*(3), 197–223.

Brown, G. M., & Johnson, R. W. (1984). Pollution control by effluent charges: It works in the federal republic of Germany, why not in the US. *Natural Resources Journal, 24*(4), 929–966.

Kant, I. (1977). *Kritik der praktischen Vernunft. Grundlegung zur Metaphysik der Sitten: Bd. VII* (W. Weischedel, ed.; 2 Auflage).

Klauer, B., Manstetten, R., Petersen, T., & Schiller, J. (2017). *Sustainability and the art of long-term thinking*. Routledge.

OECD. (1987). *OECD Environmental data, Compendium 1987.* (Vol. 42).

Roberts, M. J. (1974). *The political economy of the Clean Water Act of 1972: Why no one listened to the economists.* Prepared for the OECD.

Wieland, W. (1998). Kants Rechtsphilosophie der Urteilskraft. *Zeitschrift für philosophische Forschung, H., 1,* 1–22.

[7] The contents of this concept are based on: Faber, M., Frick, M., Zahrnt, D. (2019) MINE Website, Power of Judgement, www.nature-economy.com.

Reading Recommendations for Further Reading on Part 2 "Man and His Actions"

Baumgärtner, S., Petersen, T., & Schiller, J. (2018). The concept of responsibility: Norms, actions and their consequences. *Social Science Research Network*. [Responsibility]

Gigerenzer, G. (2015). *Bauchentscheidungen: Die Intelligenz des Unbewussten und die Macht der Intuition*. C. Bertelsmann. [Power of Judgement]

Gintis, H. (2000). Beyond Homo economicus: Evidence from experimental economics. *Ecological economics, 35*(3), 311–322. [Concept of Man]

Göpel, M. (2016). *The great mindshift: How a new economic paradigm and sustainability transformations go hand in hand*. Springer Nature. [Concepts of Man]

Jonas, H. (1979). *Das Prinzip Verantwortung. Versuch einer Ethik für die technologische Zivilisation*. Suhrkamp. [Responsibility]

Lange, P., Driessen, P. P., Sauer, A., Bornemann, B., & Burger, P. (2013). Governing towards sustainability—Conceptualizing modes of governance. *Journal of environmental policy & planning, 15*(3), 403–425. [Sustainability and Justice/ Concepts of Man]

Posner, E. A., & Weisbach, D. (2010). *Climate change justice*. Princeton University Press. [Sustainability and Justice]

Rockström, J., Gupta, J., Qin, D., Lade, S. J., Abrams, J. F., Andersen, L. S., Armstrong McKay, D. I., Bai, X., Bala, G., Bunn, S. E., Ciobanu, D., DeClerck, F., Ebi, K., Gifford, L., Gordon, C., Hasan, S., Kanie, N., Lenton, T. M., Loriani, S., & Zhang, X. (2023). Safe and just earth system boundaries. *Nature, 1–10*. https://doi.org/10.1038/s41586-023-06083-8. [Sustainability and Justice]

Schneidewind, U. (2018). *Die große Transformation: Eine Einführung in die Kunst gesellschaftlichen Wandels*. S. Fischer Verlag. [Sustainability and Justice]

Torkler, R. (2016). *Philosophische Bildung und politische Urteilskraft: Hannah Arendts Kant-Rezeption und ihre didaktische Bedeutung*. Verlag Karl Alber. [Power of Judgement]

Young, O. R. (2021). *Grand challenges of planetary governance: Global order in turbulent times*. Elgar. [Sustainability and Justice]

Part III
Time and Nature

Three Concepts of Time: Why a Clock Does Not Tell us Everything About Time and Why One Can Miss Time

8

Contents

What's it About?

It's about time. More precisely, this chapter is about different dimensions of time and why these are particularly important for dealing with environmental problems. We distinguish three dimensions of time:

1) the everyday *linear* time (Chronos), e.g. the time of day,
2) the *inherent time* of processes, e.g. natural regeneration times, and
3) the *right moment* in time *(Kairos)*, e.g. the opportune moment.

Using the example of water pollution, we demonstrate that successful environmental policy requires an adequate assessment of these three dimensions of time.

M. Faber et al., *Sustainable Action in Economy and Society*,
https://doi.org/10.1007/978-3-662-69122-9_8

8.1 Introduction to the Concept

Throughout human history, many books have been written about time, and many more will be added to this collection. Despite numerous publications about time, which is so omnipresent in people's lives, it is not easy to say what time is. It is a constant companion, it is always there, even before and after our lives, even before or after the existence of humanity. Time is intuitive on the one hand, we all have some sense of time, on the other hand, it is so complex that it is difficult to grasp, to talk or write about. Already centuries ago, the Roman bishop Augustine of Hippo (354–430 AD) expressed his difficulties in dealing with the concept of time:

> *"What then is time? If no one asks me, I know what it is; if I wish to explain it to him who asks, I do not know"* (Augustine, 1980, p. 264).

We too will not be able to grasp time in its entirety in this chapter. However, when dealing with environmental problems, it is necessary to think about time or develop a sense of time. Each of our climate goals refers to time, the drafting of (environmental) laws takes a certain amount of time, the often demanded industrial, societal and political transformation does not happen overnight, and the success or failure of movements like Fridays for Future or the Last Generation is—at least in part—dependent on the timing of the protest.

We introduce three dimensions of time that are particularly important to consider when dealing with environmental problems. These are

1. the *Chronos,*
2. the *inherent time* and
3. the *Kairos.*

On 1. The *Chronos* is the linearly progressing time, i.e., the form of time that appears most omnipresent to us in our everyday life. It is the homogeneously progressing time that we can read on a clock. Our daily routine is usually structured by it, and we plan our lives on this linearly progressing timeline, in the form of weekdays and holidays, months and years. Not only our lives, but also environmental problems, from their origin, through their discovery, their impact on humans and nature, to their possible resolution, can be dated by linear time. It may seem trivial, but only the classification of environmental problems in Chronos, i.e., on a homogeneous time scale, enables an understanding of the lengths of time over which environmental problems extend, and is thus a basic prerequisite for the development of appropriate solution strategies. For example, if we recognize that climate change is not a problem of the last 20 or 30 years, but has its beginnings in the industrial revolution over 200 years ago and is thus deeply rooted in our production and way of life, this expands our understanding of the complexity of the problem and the associated difficulties of humanity in solving the climate problem.

On 2. Every process, whether in nature or in a society, has a time inherent to it. *Inherent times* of natural processes are, for example, the approximate duration of a pregnancy, the months from sowing to harvest, or the half-life of a radioactive substance. But our societal processes too follow inherent times, such as the duration of a legislative period, the time required to switch to a more climate-friendly steel production, or the years that go by in the drafting of a new (environmental) law.

Inherent times, unlike the *Chronos*, depend on the subjective properties of the respective processes. Thus, legislative periods differ depending on the legal situation, a steel mill can only produce climate-neutral when all necessary technologies and raw materials, such as green hydrogen, are sufficiently available, and the duration of the drafting of a law is strongly dependent on political will, the scope of the law, and the legal expertise of those involved. The process of ceasing to burn fossil fuels as a country, or even as humanity, also follows inherent times. These are influenced by the factors that drive such a change, such as the recognition of the necessity of a transformation, the willingness of politics and society to transform, and the resulting climate legislation and action.

On 3. Among the many gods of ancient Greece, there was also the *Kairos,* the god of the opportune moment for a decision. The Greeks recognized that, in contrast to the linear *Chronos,* there is another qualitative dimension of time, namely singular points or periods on the chronologically progressing timeline, which are particularly favorable for certain decisions; they referred to these as *Kairos*. The *Kairos* encounters us again and again in everyday life, also in connection with environmental problems. It is always related to action. For example, Greta Thunberg hit a Kairos in the summer of 2018 when she attracted worldwide attention with her school strike and triggered a global climate movement. The conditions for her success were given—even though she might not have known it beforehand. An increased environmental awareness, communication formats that made her extraordinary protest known worldwide, as well as the accumulating signs of a globally impacting climate change, which gave her protest urgency, and much more favored the timing of her protest. We cannot know it, but it is quite possible that a similar strike ten years earlier would have led to nothing but the request to please go back to school as soon as possible.

In Summary: We need an understanding of time if we want to understand and solve environmental problems. The classification of environmental problems along the Chronos (the linearly progressing time) is important, but must be supplemented by an understanding of the inherent time periods in the relevant processes (inherent time) as well as the assessment of the right time to act (Kairos).

Important to Know: The Three Dimensions of Time

1. **Chronos:** The linear, homogeneous understanding of time, which accompanies us in our daily life. Chronos is the basis of our sciences, as it includes uniform moments and periods. Thus, an hour today is the same as an hour tomorrow.
2. **Inherent Time:** The time inherent in any given process, such as the half-life of a radioactive substance or the time required to draft a law. Unlike Chronos, the inherent time of a process depends on various factors influencing the process and is therefore often difficult to estimate in advance. For example, a legislative process can take years or even decades, depending on the technical and social context.
3. **Kairos:** The right time to act. Kairos fundamentally differs from Chronos, as it is neither continuous nor can it be divided into homogeneous units. Different times and moments differ qualitatively in this view, a climate law can be passed today rather than 30 years ago. The right time, or the right period to act can be linked to Chronos by dating it on a timeline.

8.2 Time as a Crucial Variable for the Complexity and Solution of Environmental Problems

The fact that environmental problems are not static, but always arise, have effects and are solved over periods of time, has already become clear in the above explanations. To deal appropriately with environmental problems, i.e., to resolve them or better yet not let them arise at all, decision-makers in society should consider all three dimensions of time—Chronos, inherent time, and Kairos. If this happens and the dimensions of time are well estimated, it can contribute significantly to the solution of environmental problems, as we will explain in the case of German wastewater legislation in the 1970s. Conversely, if the time dimensions of processes or problems are poorly estimated, this can even lead to the failure of otherwise well-thought-out solution approaches, as the wastewater legislation of the USA shows, which was passed in the same decade.

8.2.1 Wastewater Legislation in the USA[1]

In 1972, the *Clean Water Act* came into force in the USA. The largely industrially caused pollution of surface waters was to be largely freed from pollution within 13 years by this ambitious law. The fact that the implementation of the law was ultimately not crowned with success had mainly two reasons. On the one hand,

[1] The following explanations are based on Klauer et al., (2013, pp. 130–135).

there was strong resistance to the wastewater regulations from powerful industrial interest groups. Although this resistance could probably have been resolved, it was mainly the second, temporal reason that caused the law to fail. The implementation of the regulations, i.e., a complete protection of the waters through the cleaning, reduction, and avoidance of wastewater, was not possible for the companies within the tight deadlines, as the necessary technologies were not yet available. The inherent time of the required industrial transformation was underestimated.[2]

8.2.2 Wastewater Legislation in Germany

Less ambitious (in terms of time and with regard to the reduction of the quantity and degree of wastewater pollution) than the American water legislation was the Water Management Act passed in Germany a few years later (1976), within the framework of which the Wastewater Charges Act was also adopted. The law included a monetary levy from companies that discharged polluting wastewater into bodies of water. However, this levy was not due until five years after the law was passed and was initially relatively low, but was to be gradually increased over time. The law was heavily criticized by business associations, environmental associations, and environmental economists for various reasons, primarily due to the relatively low and delayed levies on wastewater—especially compared to the USA. On the one hand, businesses laimed the cost burden to be ruinous, on the other hand, environmentalists criticized that the emitters of wastewater still did not pay the real costs of the pollution. In their view, the wastewater levy was initially much too low and came into effect too late. In addition, the levies incurred should also be paid, and not instead be offset by investments in wastewater treatment plants.

Contrary to the expectations of the economy and environmental economists and associations, the supposedly devastating or insufficiently ambitious law was successful. Water pollution was reduced more than expected. In Baden-Württemberg, for example, the industry achieved an emission reduction of 40% between 1975 and 1979, even 2 years before the introduction of the water levy (Faber et al., 1989, p. 56). This success was achieved through great efforts by the authorities and industry, who invested large sums in the construction of wastewater treatment plants in order to pay as little levies as possible in the base year 1981. The German law was subsequently adopted in a modified form by many other countries.

---

[2] In addition to the short period of 13 years, the objective of completely avoiding all wastewater was problematic, as this cannot be implemented. For thermodynamic reasons, the absolute reduction of all wastewater to zero is only possible if all production and consumption is stopped (see Chap. 9 *Thermodynamics*).

8.2.3 Success and Failure Due to Different Assessments of Time

Why did the "ambitious" law fail in the USA and why did the rather lax wastewater law in Germany succeed? A crucial reason lies in the fact that the legislators of both countries assessed the inherent times of the affected processes differently.

The goal of both laws was to significantly reduce and clean industrial wastewater to ensure clean waters. To achieve this goal, an industrial transformation process was necessary, in which wastewater treatment plants and wastewater systems had to be built, dirty technologies had to be made more efficient or replaced by cleaner technologies. Also, because some of the required technologies had to be developed first, this process—like similar processes today—took a period of time, the length of which could not be determined exactly in advance due to the many uncertainties (see Chap. 12 *Ignorance*).

The US wastewater law gave polluting industries 13 years to complete the necessary transformation process. A period that would not have been sufficient under the given circumstances, neither in the USA nor in Germany, to reduce the pollution to the ambitious targeted level, which would have allowed swimming and fishing in the waters. The consequence in the USA was that the companies made no or only few efforts to reduce their wastewater and the law had to be fundamentally rewritten a few years later. As a result, important years for environmental protection were lost.

In Germany, the inherent time of the transformation process was estimated to be longer, which contributed to the good result of the law. Only after 5 years was the first, relatively low levy on wastewater due, which then gradually increased. Unlike the case in the USA, the economy did not face insurmountable challenges, but was given enough time to act decisively and transform—ultimately even faster than expected. If the German law had been stricter and formulated with shorter periods—i.e., an underestimation of the inherent time—it might have led to similar problems as in the USA.

This example shows how crucial the assessment of the inherent times of transformation processes is. Timeframes that are estimated too short can lead to attempts to change not being made at all in the face of hopelessness. However, this does not mean that timeframes can be chosen arbitrarily long. Too long timeframes could lead to nothing happening for a long time, under the motto "there is still plenty of time to do something", and that existing environmental problems continue to exist or worsen, even though they urgently need to be solved.

8.3 Three Questions about Time: A Checklist for Problem Solving

Time plays a crucial role in all environmental problems. No matter what the topic is, time is always there, it accompanies all environmental problems and significantly influences them and especially possible solutions. The problem is that the causes and effects of environmental problems often take up so much space in political and societal debates that the temporal aspects are pushed into the background and are sometimes little noticed, sometimes simply ignored. Then the solutions that emerge from the debates may be well-intentioned in principle, but they do not achieve their goal if, for example, they are not feasible in the envisaged time frames, if no opportune moment is chosen for their implementation, or if they are simply implemented too slowly.

An assessment of the three dimensions of time is a measure that is worthwhile for all people who deal with environmental problems. This means that it is never wasted effort when we fundamentally ask ourselves three questions about time when dealing with environmental problems:

- What is the chronological timeline of the problem? *(Chronos)*
- What are the inherent times of the processes relevant to the problem? *(Inherent time)*
- Are there times that are particularly favorable for solving the problem? *(Kairos)*

These three questions provide a checklist that can be referred back to throughout the process of solving an environmental problem in order to classify the temporal background of the problem.

However, it is not the case that these questions can be answered once at the beginning and these answers then have permanent validity. Especially temporal considerations are associated with a lot of *ignorance*, which is partly irreducible (see Chap. 12). For example, we cannot know the exact Chronos of the future and even the Chronos of the past often causes problems for historians. The inherent times of many processes can be estimated, but new developments, for example of a technological nature, can nullify the initial estimates. Also, the various favorable times in the timeline are difficult or even impossible to determine in advance.

Therefore, when answering the checklist, we are dependent on our *Power of Judgement* (see Chap. 7). On the one hand, it is important to combine existing knowledge with experience to give good answers. On the other hand, the application of judgement also includes recognizing when we cannot (yet) answer part of the checklist and perhaps do not yet need to answer it. This prevents us from "wasting our time unnecessarily thinking about time". The checklist is therefore an iterative tool, which means that we keep going back to it during the solution process, orient ourselves to previous answers, adjust these if necessary to our advanced knowledge, and add answers where we could not give any before. This makes the checklist a temporal orientation aid for environmental problems that everyone can always keep in mind to better deal with the multitude of environmental problems we face every day.

8.4 Time as the "Final Boss" of Wishful Thinking

When developing solutions for an environmental problem, not only the theoretically possible effect of a solution is important, but above all whether the solution can be implemented, or within what time frame and when it can be implemented. For example, from an environmental and energy policy perspective in Germany, it would be desirable if a large proportion of buildings previously heated with natural gas had been converted to electric heat pumps as early as November 2022, when the Russian attack on Ukraine led to a severe shortage of natural gas supply in Germany. If in addition the electricity required by these heat pumps was generated renewably, this "solution" would have brought Germany closer to its climate goals, besides making it more independent of gas imports from Russia and other, politically problematic energy exporting countries. However desirable a rapid implementation of such a solution may be, it is not feasible within a few months. A transformation process of this magnitude, in which millions of new heating systems have to be installed in existing and new buildings, electricity generation capacities have to be massively expanded through wind and solar, and additional energy storage options have to be created for bridging periods of calm, takes years, possibly even decades.

While the length of this period also depends on political and societal will, even with a determined will to transform, the period cannot be arbitrarily shortened. In addition to gigantic amounts of materials needed for the construction of heat pumps and renewable energy generation and storage facilities, which are difficult to obtain in the short term, the new facilities have to be manufactured and—perhaps even more crucially—subsequently installed. This requires a large number of skilled workers who need to be trained over years (see Chap. 15 *Stocks*). In the long term, all this is possible and, according to current knowledge, also sensible, but in the short term, this desirable, overarching transformation can neither reduce the climate problem nor Germany's dependence on gas and oil exporting countries and therefore needs to be supplemented by shorter-term, less ambitious solutions.

This example illustrates that reflecting on time can help to better assess solution strategies, particularly their implementation. When we think about inherent time, for example, we understand why Robert Habeck—a member of the Green Party and Minister for Economy and Climate Protection—felt compelled to travel to Qatar in the fall of 2022 to negotiate the supply of climate-damaging liquefied natural gas (LNG), even though he and his ministry were aware that this could not be an optimal solution for either the climate problem or the issue of Germany's energy import dependency.

Staying with this example, we also see the importance of Kairos for the implementation of solution approaches. Despite all the negative effects of the energy crisis in Europe in 2022, triggered by the Russian attack on Ukraine, the crisis could represent a *Kairos* for the energy transition. For instance, sharply rising prices for natural gas made the electricity-based use of heat pumps economically more competitive compared to then significantly more expensive heating with natural gas: a window of opportunity (Kairos) to accelerate the expansion of climate-friendly technology. Furthermore, the energy policy necessity to become more independent from Russian gas could act as a driver for an accelerated expansion of renewable energies, as society recognized the need for expansion and was more willing to accept higher costs and other restrictions than before the crisis. Politicians, business representatives, and scientists who recognized the crisis as a Kairos for an energy transformation, despite all the problems, could contribute to setting the right course for an accelerated and thus also more climate-friendly energy transition.

In addition to assessing the length of transformation processes and the favorable times for them, reflecting on time also gives us guidance on how long these transformation processes should last at most. Because not only solution approaches are determined by time, but also the progressing environmental destruction leading to problems such as climate change, species extinction, pollution of the seas, and others. The natural cycles affected by this are partially irreversibly thrown out of balance with progressing environmental destruction (see Chap. 10 *Irreversibility*). Therefore, we cannot afford to take unlimited time with either the German or the global energy transition. Because even if the scientific forecasts do not provide us with exact time frames, we can still recognize on their basis that a too slowly implemented transformation towards environmentally and climate-friendly actions and production methods will lead to irreversible changes with devastating consequences for human life on Earth.

The inherent times thus become another filter for goal-oriented solution strategies. Expressed using the example of the energy transition or climate change: As humanity, we can no longer afford to wait another 30 years for a supposedly "saving" technology like nuclear fusion, as we would exceed the inherent times of nature beforehand and would damage our natural bases of life so massively that human life on Earth, as we know it today, would only be possible to a very limited extent.

Summary: The Importance of Time for Solution Strategies in Environmental Issues

When searching for suitable solution strategies, time plays a crucial role. It is important to assess time both in direct relation to the solution strategy and in relation to the natural characteristics of the environmental problem itself. For this, it is helpful to ask:

- How much time does the implementation of a certain solution take, and are there certain favorable times for the implementation of the strategy? (Strategy-specific)
- What are the temporal conditions of the environmental problem? E.g., tipping points in climate change, inherent time of recovery of an ecosystem, etc. (Nature-specific)

Temporal considerations in these two respects can provide guidance on which solution strategies are feasible at which time and which are not.

References[3]

Saint Augustine (1961). *Confessions. Translation R.S. Pine-Coffin*. Penguin Classics.

Faber, M., Stephan, G., & Michaelis, P. (1989). *Umdenken in der Abfallwirtschaft: Vermeiden, Verwerten, Beseitigen*. Springer.

Klauer, B., Manstetten, R., Petersen, T., & Schiller, J. (2017). *Sustainability and the Art of Long-Term Thinking*. Routledge).

[3]The contents of this concept are based on: Klauer et al. (2013) Chap. 8 and Faber, M., Frick, M., Zahrnt, D. (2019) MINE Website, Time, www.nature-economy.com.

Thermodynamics: Fundamentals about the Nature of Environmental Problems

9

Contents

What's it About?

It's about energy. Every economic activity requires energy. The use of energy always means that energy is being converted. This conversion follows physical laws, which are formulated with the two main laws of thermodynamics. To understand why environmental problems

---

This chapter was first published in a slightly adapted form as a discussion paper: Faber, M., Rudolf, M., Frick, M. & Becker, M.-Y. (2023). Thermodynamik - grundlegende Einsichten für ein Verständnis von Umweltproblemen, AWI Discussion Paper Series No. 725, Heidelberg.

M. Faber et al., *Sustainable Action in Economy and Society*,
https://doi.org/10.1007/978-3-662-69122-9_9

necessarily arise from our industrial economic system, we must deal with thermodynamics, that is, with the area of physics that considers the laws of energy.

This chapter will first introduce the basic concepts and laws of thermodynamics. We will then show how thermodynamics contributes to a better understanding of both the emergence and the elimination of environmental problems.

9.1 Introduction to the Concept of Thermodynamics

The considerations of this chapter start with the concern to find a scientific approach to environmental problems that arise from human activity, especially from economic activity.

For this, it is important to first clarify why economic activities lead to environmental problems at all. This question is currently not receiving the attention it needs. Either the causal link between economic activity and its negative effects on the environment around us is accepted and not further questioned, or the analysis quickly focuses on analyzing the specific processes that are at work in individual cases—for example, regarding the pollution of a river.

We want to show a different way in the following, which allows us to make fundamental statements about how economic action necessarily has effects on the environment due to scientific interrelationships, and why these effects are often harmful. This is particularly true for human action in the industrialized economy.[1] We want to show a way that enables us to easily recognize when economic action is harmful to the environment.

In order to systematically understand the necessary connection between human action and its effects on the environment, we will consider whether it is not possible to find an influence that is not only important in all economic actions, but even in all natural processes. Could it perhaps be that such an influence exists and is also a major cause of the emergence of environmental problems?

Indeed, there is such an influence, namely energy. It is easy to see that it is not possible to achieve anything without the use of energy, let alone carry out a useful economic action. The basic consideration of the influence of energy on human activities will help us find ecological and economic foundations that allow a comprehensive view of the environment and the economy. It allows us, including the following considerations, to find appropriate solutions to environmental problems.

[1] Even in antiquity, the philosopher Plato complained about the overexploitation of wood in the Mediterranean forests for the construction of fleets. However, we know from the writings on ethics from ancient Greece that human influences on nature were largely unknown and therefore hardly reflected ethically.

9.2 Different Forms of Energy and the Concept of Thermodynamics

In the first step of the scientific consideration of energy, we find that there are different forms of energy. Particularly important are:

- kinetic energy (possessed by a body due to its speed),
- potential energy[2] (energy of position; this is the basis for hydroelectric power plants, which are driven by the water from dams),
- electrical energy,
- chemical energy,
- nuclear energy,
- radiation energy (electromagnetic waves, solar energy) and
- thermal energy.

Of all the forms of energy mentioned, thermal energy has a special significance, because every other form of energy can never appear isolated, but only together with thermal energy. For example, chemical energy appears as fire, motion energy is associated with friction, which leads to heat. For this reason, the branch of physics that primarily deals with energy is called *thermodynamics*; the ancient Greek word *thermós* meaning 'warm' and the ancient Greek word *dýnamis* meaning 'force'. In German, thermodynamics is therefore also referred to as the study of heat. However, thermodynamics is not only about heat, but also about other things, e.g. the mixing of substances.

Important to Know: Significance of Thermodynamics for Environmental Problems

Thermodynamics forms the link between our actions and their environmental impacts. As long as we have understood the basic insights of thermodynamics, we also have access to the question of how environmental problems arise. At the same time—and this is even more important for us—we thereby gain an orientation to anticipate the emergence of environmental problems: namely where energy is used. With this knowledge, environmental damage cannot always be avoided, but it can be drastically reduced.

Now, thermodynamics is not an easy field of physics. One of the most influential thermodynamicists of the 20th century, Arnold Sommerfeld (1861–1951), described thermodynamics with the following words: "Thermodynamics is a funny subject. The first time you deal with it, you don't understand anything about it. The second time you work through it, you think you have understood everything, except

[2] In physics, kinetic and potential energy are combined under the term mechanical energy.

for one or two small details. The third time you work through the material, you realize that you hardly understand anything about it, but by then you have become so used to it that it no longer bothers you" (Lauth & Kowalczyk, 2015).

Therefore, we will develop a clear representation in the following and only delve as deep as necessary for a systematic view of environmental problems. However, there are also limits to the simplicity of a good representation. Albert Einstein (1879–1955) aptly formulated this: "Things should be made as simple as possible, but not simpler." The reason for this limitation is that too much simplification can lead to a false representation. This has often happened with thermodynamics. Therefore, we will follow Einstein's advice in the following and try to strike a good balance between simplicity and complexity.

We first explain the concept of thermal energy and clarify the relationship between heat and mechanical work. This gives us the necessary basic knowledge to understand the so-called first two laws of thermodynamics.[3] After that, we turn to the difference between free and no longer usable energy, thereby creating access to the concept of *entropy*, which is central in thermodynamics but unfamiliar to many people.

9.2.1 Thermal Energy[4]

Heat is familiar to all of us. Sometimes it's too warm, sometimes too cold, sometimes the temperature is just right. This everyday observation leads us to one of the three essential properties of thermal energy, namely temperature. Heat is determined by

- the temperature,
- the amount of heat, (it is determined by its energy content, e.g. the amount of burned coal in a stove)
- and the above-mentioned entropy, which we will discuss in more detail below.

However, these essential properties of heat do not yet tell us what heat actually is. It took a long time to find out that heat is the kinetic energy of particles. What does this mean and how was it discovered? Some may still know the old-fashioned mercury thermometers. The higher the temperature, the more the mercury in the glass tube of the thermometer expanded. The temperature could then be read from the height of the mercury. This observation applies not only to mercury, but—albeit to varying degrees—to many materials that expand with increasing temperature.[5] The reason for this is that every molecule, indeed every atom, is constantly

[3] Some famous physicists, like Arthur S. Eddington (1882–1944), even considered the 2nd law of thermodynamics to be the most important natural law of all.

[4] See for the following Hüfner & Löhken (2010, pp. 253–294).

[5] There are exceptions. For example, ice floats on water, i.e., ice is more expanded than water.

in motion. When heat is added, the thermal energy of the system increases; because the individual atoms or molecules that make up the system move faster, i.e. their kinetic energy increases. The greater the thermal energy of the particles, the faster the movement. Only at absolute zero, i.e. at minus 273 degrees Celsius, does the movement of the particles come to a standstill.

9.2.2 Heat and Physical Work

What is the relationship between heat and physical work?[6] This relationship can be illustrated using the example of steam locomotives, which are now only known from films. Often a stoker was used there, who heated the steam boiler with coal. The steam generated by coal combustion—i.e. with heat—drives a piston engine through the pressure it creates; this causes the train to move through its kinetic energy; physically speaking, work (force times distance) is thus performed.

The following example illustrates how concentrated the energy in the form of thermal energy is: Let's consider a pot of water and heat the water by 1 degree Celsius. Now we can ask ourselves how many meters the pot could be lifted if the same amount of energy that was needed for heating was used in the form of work? Even some physicists would be surprised that the answer is: 430 m! (Hüfner & Löhken, 2010, p. 266).

Equal amounts of energy are, economically speaking, worth different amounts: For example, the price of 1 kWh of electrical energy is significantly more expensive than 1 kWh in the form of heat, which was produced, for example, by natural gas. The production costs of electrical energy are about 2 to 3 times as high as those of thermal energy with the same heating material. At first glance, this is surprising; because both amounts of energy do the same when heating. Why do these two forms differ so much economically? One clue is that electrical energy, like all other forms of energy, can be converted into heat almost without loss. In contrast, heat can never be completely converted into electrical energy, nor into all other forms of energy. Thermal energy is therefore fundamentally different from all other forms of energy. It can only be transported or converted into other forms of energy with great energy losses. It is therefore only partially usable for human purposes. This has a number of consequences for the economy and the environment.

The reason for the limited usability of thermal energy results from the two main laws of thermodynamics.

---

[6] Physical work = force (F) times distance or path (s) = $F \cdot s$. The force F acts on a body, which travels the distance s in the direction of this force.

9.3 The Two Main Laws of Thermodynamics: Energy Conservation and Entropy Law

9.3.1 The 1st Law

To understand the 1st law, it is necessary to explain the concept of the *thermodynamic system*. Thermodynamic systems are spatially delimited and differ by three characteristics:

1. *Isolated systems* do not exchange energy or matter with their environment.—Example: A closed cooler (in a good approximation).
2. *Closed systems* exchange energy, but not matter, with their environment.—Example: Earth (except for comets and rockets etc.). Heat exchange with the universe occurs in the form of solar radiation energy.
3. *Open systems* exchange both energy and matter with their environment.—Example: humans. We take in matter through eating and drinking and release matter in the form of excrement, and exchange thermal energy with the environment.

Whether there are any natural systems that exactly correspond to these distinctions is, however, an open question. Real systems on Earth always exchange—at least in small amounts—energy and matter with their environment.[7]

What is outside a system, we call its "environment". Whenever we use the word "system" in the following, we mean this thermodynamic definition: the relationship of a delimited "inside" to a surrounding "outside". The boundaries of a system are always determined by what the interest is focused on, and thus includes the interactions of the system elements encompassed (inside). We can understand a steam locomotive as a system if we are interested in the processes and interactions of a steam locomotive with its environment, the outside, (e.g., emissions into the atmosphere). However, if our interest is in the processes and interaction of the elements in the steam boiler (e.g., the behavior of the water molecules when heat is supplied), then the system boundary is to be chosen so that the steam boiler is defined as the system, whereby the rest of the steam locomotive and the further environment are defined as "outside". Simplified, a system can be represented as in Fig. 9.1.

Now we can formulate the 1st law of thermodynamics. It states:

In an isolated system, the total energy and the total mass are always constant.

This statement may be surprising at first glance. How can it be that the energy contained in a candle, after it has completely burned out and extinguished, is still

[7] If the period under consideration is not large, then we can neglect this exchange due to the small amount.

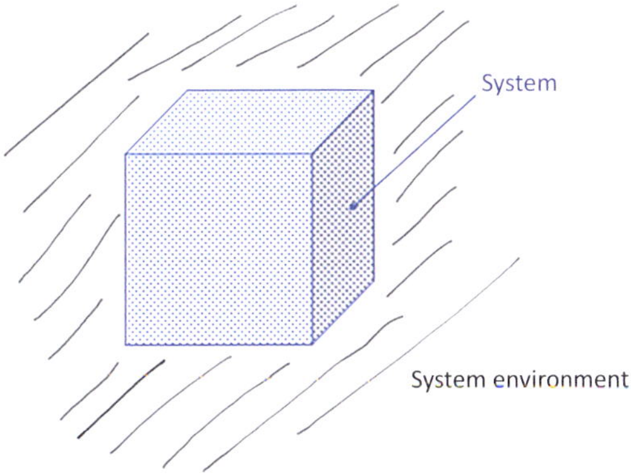

Fig. 9.1 Schematic representation of a thermodynamic system and its environment (Source: Own illustration)

preserved? Let's assume it was in an isolated system. Then the chemical energy of the candle has turned into heat during combustion, accordingly, the heat in the isolated system has increased by the same amount. Consequently, the total energy, which is equal to the sum of chemical energy and thermal energy, has remained the same according to the 1st law. However, physically speaking, a lot has happened in our system; what has happened is explained by the *2nd law*.

9.3.2 The 2nd Law and the Concept of Entropy

The 2nd law says something about how the energy changes *qualitatively* during an energy transformation. It expresses the direction in which energy transformation is possible. For example, it is possible to completely convert mechanical, electrical, or chemical energy into *thermal energy*. However, as mentioned above, thermal energy can only be partially converted into the aforementioned energies.

So how can one determine the possibilities for the qualitative change of thermal energy? Each change of an energy form is determined from the product of an intensive variable with an extensive variable. How do these two terms differ? An extension is an expansion. Intensive, on the other hand, is referred to as "penetrating" in relation to sensory impressions. We want to explain this using examples. Let's consider two bodies, one weighing 60 kg, the other 40 kg. The weights are then the extensive variables. Let's assume both are moving at a speed of 200 kilometers per hour. The speed is then the intensive variable. Let's further assume that both bodies unite in motion. Then the extensive variable of the united body is the sum of the two extensive variables, i.e., 100 kg, the intensive variable, however, remains constant at 200 kilometers per hour. The same applies if we consider the two bodies and this time heat them to a temperature of 30 degrees Celsius. The intensive variable of each of the two bodies is their temperature. If we again unite them, then the extensive variable, the weight, increases again to 100 kg, the intensive variable, however, remains constant at 30 degrees.

Let's now consider various forms of energy. For example, work results from the multiplication of the intensive variable force with the extensive variable distance. In the case of electrical energy, the intensive variable is voltage and the extensive variable is current. With gas pressure, the intensive variable is pressure and the extensive variable is volume. Let's now turn to thermal energy. Obviously, temperature is the intensive variable. But what could be the extensive variable? While for all other forms of energy the extensive variable is easily observable in reality, no observable size could be found for the extensive variable of thermal energy, although numerous scientists have long dealt with this topic. Finally, the physicist Rudolph Clausius (1822–1888) had the brilliant idea to introduce a non-observable extensive variable to describe the change in thermal energy; he called it *entropy*. This word is derived from the ancient Greek term entropía, which means 'turning'. A change in thermal energy thus results as the product of the intensive variable, temperature, with the extensive variable, the change in entropy (Fig. 9.2).

From Fig. 9.2, it also follows that during a process, the change in entropy is influenced by the temperature (Fig. 9.3).

Any change in the amount of heat thus causes a change in entropy. However, the magnitude of the change in entropy is smaller the higher the temperature and larger the lower the temperature of the entire process.

At this point, it already becomes clear that entropy is not an easily understandable quantity. The physicist Max Planck (1857–1947) once said: "Entropy is something you don't understand, but you get used to it" (Hüfner & Löhken, 2010, p. 272, our translation). It is difficult to imagine something tangible under it. Now that we have become a little accustomed to entropy, we can state the 2nd law of thermodynamics. It reads in the formulation of Rudolf Clausius (1822–1888) from the year 1865:

In an isolated system, entropy cannot decrease.

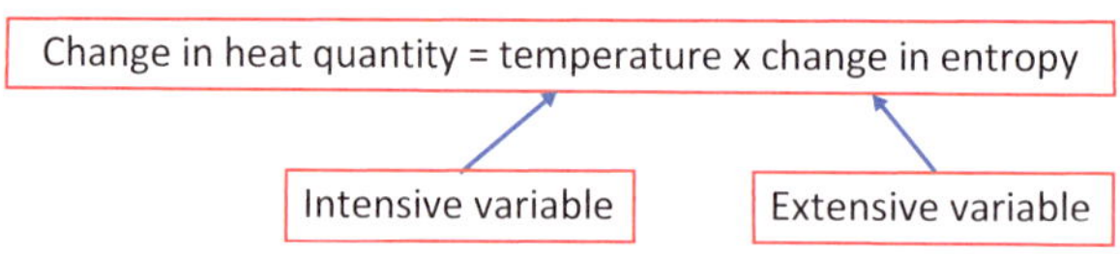

Fig. 9.2 Change in thermal energy as the product of the intensive variable "temperature" with the extensive variable, the "change in entropy"

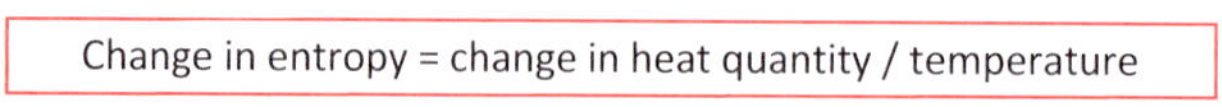

Fig. 9.3 Change in entropy is influenced by temperature (according to the physicist Rudolph Clausius, 1822–1888)

Therefore, the creation of entropy always involves a change in one direction. Entropy can only arise by itself, but cannot be destroyed.

The 2nd law of thermodynamics has far-reaching consequences. From school lessons, some may still remember the following conclusion from the 2nd law: A perpetual motion machine is not possible, because energy is consumed by the movement. For example, even a pendulum will eventually come to a standstill, as inevitably kinetic energy is converted into thermal energy through friction with the air.[8] Once entropy in an isolated system has reached its maximum, the system is in a stable equilibrium state: all change processes, e.g. temperature changes, have ended.

This insight shows: Certain processes are irreversible. This means that processes are not reversible, such as the burning of a candle (see Chap. 10 *Irreversibility*). Even if the candle were to burn in an isolated system and thus all energy is converted into thermal energy and remains without losses, it is not possible to restore the candle. This is because entropy is created during the combustion process. Another example of irreversibility is an iron rod that is hot at one end and cold at the other. After some time, the iron rod will have the same temperature at all points. However, it will never happen that one end becomes hot again and the other cold by itself. This irreversibility is so important for environmental problems that we discuss it in detail in various scientific publications and within the framework of our project *MINE—Mapping the Interplay between Nature and Economy* (www.nature-economy.com) in a separate concept *Irreversibility* (see Chap. 10 and Faber et al., 1996, pp. 103–112, 1998, pp. 74–82).

9.3.3 Free Energy and Entropy

The second law allows us to determine how much of one type of energy can be converted into physical work. For this purpose, the term "free energy" was introduced in physics. The definition equation for free energy F is

$$F = U - T \times S,$$

where U is the total energy of the system, S its entropy, and T the (absolute[9]) temperature. T x S is the product of temperature and entropy. The free energy is the portion of the total energy that can be converted into work. The product of temperature and entropy (T x S) is the portion of the total energy that cannot be converted into work. The second law tells us that the free energy in a system cannot increase, but can only decrease through energy conversions.

[8] For a detailed proof see Faber, Niemes & Stephan (1983, pp. 88–90).

[9] On the Kelvin temperature scale, starting at the absolute zero point of -273 ° Celsius.

Important to Know: The Two Main Laws of Thermodynamics

- **The first law states:** *In an isolated system, the total energy and the total mass are always constant.* It therefore says something about the conservation of mass and energy.
- **The second law states:** *In an isolated system, entropy cannot decrease.* It says something about how forms of energy change. This is possible with the concept of entropy. How this can be illustrated is shown below.

9.3.4 Degree of Order as an Illustration of Entropy

As mentioned above, it is not possible to imagine something tangible under entropy. However, it is possible to imagine a quantity that is directly correlated with entropy; this was shown by the physicist Boltzmann (1844–1906): The higher the entropy of a system, the greater the disorder in the system. For example, ice has an ordered structure of ice crystals; if it melts due to energy input in the form of heat, then water molecules are formed, which are obviously much more disordered due to their movement: The entropy of the system, the degree of disorder, has increased.

Important to Know: Illustration of Entropy

- If entropy increases, then the degree of order decreases, meaning: The (molecular) disorder increases.
- This circumstance leads to the fact that in many cases the increase in entropy is something negative for humans, as systems with lower order contain less free energy—thus usable energy.

Thermodynamic relationships have many economic consequences. This insight was particularly formulated in the context of Ecological Economics under the term *joint production* (See Chap. 13 and Baumgärtner et al., 2006). Joint products are by-products of production processes that necessarily occur thermodynamically, are usually unwanted, and in many cases lead to environmental damage. Prominent examples of environmental damage caused by joint products are the climate problem, the loss of biodiversity, and the global water problem.

Due to the difficulty of thermodynamics, especially entropy, we recommend the 12-minute video contribution by Martin Buchholz to the final of the German Science Slam Championship. Using the example of cooling towers, Buchholz manages to provide a well-understandable, illustrative, and even humorous introduction to thermodynamics and the concept of entropy. The contribution can be found on the YouTube platform under the search terms *"Kühltürme"* and *"Science Slam"* or directly via the link given in the footnote.[10]

9.4 Considering Thermodynamics in Economics

In the first half of the 19th century, physics experienced a revolution with the development of the field of thermodynamics. Phenomena related to energy were explored both theoretically and practically. The motivation behind this research was economically driven, as the aim was to improve the efficiency of the Watt steam engine.

However, thermodynamic considerations only found their way into economic studies in the last quarter of the 20th century. The Romanian economist Nicholas Georgescu-Roegen (1906–1994) was the first to focus his investigation of the economy on energetic considerations in his groundbreaking work *The Entropy Law and the Economic Process* (1971). He was able to show how closely intertwined economy and environment are due to thermodynamic relationships.

It is actually strange that one of the most important, perhaps even the most important factor of production in the economy, namely energy, did not have a prominent role in mainstream economics, whose foundations were laid in the last quarter of the 19th century. Although energy is one of the most important economic factors of production, thermodynamics still does not play a prominent role in large parts of economics today. However, energy is necessary for every production process and its use has effects on nature. It leads to an irreversible conversion of fossil fuels such as coal, oil, and gas into heat, CO_2, and other greenhouse gases as well as pollutants (e.g., fine dust), thus contributing to the climate problem and various other environmental problems. In particular, Georgescu-Roegen highlighted the negative side of industrial production. On the one hand, an energetic raw material, such as coal, oil, or gas, cannot be used energetically again after its use. This process is therefore irreversible (see Chap. 10 *Irreversibility*). Perhaps even more important, however, is that industry always produces more than is actually intended and needed—joint production and often associated environmentally harmful joint products (see Chap. 13 *Joint Production*) are inevitable (Baumgärtner et al., 2006, Chap. 3).

This fact is easy to convey and sharpens awareness of the risks of our production methods. An example is the production of steel using coke and iron ore. Not only steel is produced, but also the residues of the production process, such as CO_2, wastewater, dust, etc. These insights gained on the basis of thermodynamic relationships are central to understanding environmental problems.

[10] https://www.youtube.com/watch?v=z64PJwXy--8.

9.5 The Transition from Classical Mechanics to Thermodynamics in the Course of Industrialization

Classical mechanics deals with systems comprising of few elements, e.g., the bodies of a planetary system. They are clear and can therefore be easily described. A key characteristic is that their processes are reversible in time; a planetary system could also move in the opposite direction. This means that processes in classical mechanical systems are reversible in time. We also recognize that temperature, heat, and therefore entropy play no role in this.

The situation is quite different with complex systems, e.g., a lake consisting of many water droplets and even more water molecules. The complexity arises not only from the large number of elements in such a system, but also from the fact that these elements, in this case the water droplets, interact with each other. For example, if a lake is heated on one side by sunlight, this heat spreads throughout the lake. How is it possible to deal with such complex issues? In addition, unlike classical mechanical systems, the processes in complex systems are irreversible, i.e., not reversible: a consequence of the 2nd law of thermodynamics.

At the beginning of the 19th century, practitioners, engineers, and scientists such as James Watt (1736–1819), Sadi Carnot (1796–1832), James Prescott Joule (1818–1889), Rudolph Clausius (1822–1888), and William Thompson (later Lord Kelvin, 1824–1907) began to study such complex systems in their efforts to improve steam engines. The simplified view of classical mechanics was not sufficient for their research, so their aim from the start was to understand the interaction of natural systems with technical systems—which are created and managed by purposeful human action—in their complexity. This was an unusual approach at the time in a traditional science like physics.

9.6 The Limits of Thermodynamic Efficiency

How can engineers, technicians, or entrepreneurs, both in the past and today, practically apply their thermodynamic knowledge? A key application is the calculation of the thermodynamic efficiency of a machine. Whenever energy is used, thermal energy is generated due to the 2nd law of thermodynamics, which can no longer be used.

Important to Know: Thermodynamic Efficiency
The efficiency indicates how much of the energy used in a system is converted into the desired effect.

Let's consider two cars, one powered by a combustion engine, the other by an electric motor. The combustion engine, with average driving behavior in Germany, converts about 20% of the energy contained in the gasoline into the car's motion (20% efficiency), while the electrically powered vehicle has an efficiency of around 80%. When losses are included, which occur during battery charging and power supply, the efficiency is still significantly higher than that of the combustion engine at 64% (BMVU, 2021).

Efficiency also plays a crucial role in power plants. Coal-fired power plants have an efficiency of up to 45%. The varying efficiencies of power plants can be seen, for example, in the fact that the gas and steam combined power plant (GuD) Haveli, which was built in Pakistan in 2018, had the highest efficiency of all fossil fuel power plants worldwide at 62.4% (Tractebel Engie, 2018).

The fact that thermodynamic efficiency can be greatly improved through technological progress is already evident from its first application in the case of the steam engine (Hüfner & Löhken, 2010, p. 270). The first steam engines developed by James Watt had an efficiency of only 1 percent; 99% of the energy was therefore unused. It took a long time to increase the efficiency to the current typical 30 to 50%. However, there is no escape from the 2nd law. Since entropy and thus thermal energy, which can no longer be used, is generated due to thermodynamic circumstances, the theoretically maximum efficiency is not only less than 1, but often considerably less than 1.

9.7 The Increasing Importance of Thermodynamics for Economy and Environment

We deal so extensively with thermodynamics because it forms the basis of all processes in nature and thus also in the economy. Georgescu-Roegen recognized that both biological and economic systems are open systems, i.e., they use matter and energy, absorb these from their environment, and release them back after use. The economy extracts raw materials, whether they are plant, animal, or mineral; after their use, they are released back into the environment, regularly in a harmful manner since the industrial age, i.e., since the mid-18th century. Georgescu-Roegen thus laid the foundations for a bioeconomic theory by comprehensively applying thermodynamics to the relationships between the environment and the economy. As the title of his book *The Entropy Law and the Economic Process* (Georgescu-Roegen, 1971) shows, the concept of entropy plays a decisive role. He introduced the concept of entropy into economics, which—when considered—leads to a new view of economic processes. From his considerations, it follows that increasing resource use leads to an exhaustion of the Earth's capacities, both in terms of raw materials and the environment's capacities to absorb pollutants and eliminate them: "The significant fact for the economist is that the new science of thermodynamics began as physics of economic value and, basically, can still be regarded as such. The Entropy Law [i.e., the second law of thermodynamics] itself emerges as the most economic in nature of all natural laws" (Georgescu-Roegen, 1971, p. 3). This insight led him to call for a radical new structuring of the conceptual foundations of economics. This led to the founding of Ecological Economics in 1989: Over time, various directions developed, such as the Barcelona School of Ecological Economics, the Heidelberg School of Ecological Economics, the Degrowth Movement, and many more.

9.8 Summary: Thermodynamic Fundamentals for Environmental Policy

How can central insights for the fundamentals of environmental policy be summarized? Georgescu-Roegen generally showed the consequences for the environment of the increasing amounts of highly concentrated (with low entropy)—and thus easily available—resources being used in the economy since the industrial revolution in the 18th century. For example, petroleum, which is present in the earth in high concentration (low entropy), serves as an energy source to produce high-quality materials, e.g., aluminum (low entropy), from other materials that had been extracted with low concentration, e.g., bauxite (high entropy). During production, wastewater, waste heat, CO_2, slag, fine dust, etc. (with high entropy) are produced. Also, the goods produced (low entropy) become waste (high entropy) as they degrade over time, which is then released into the environment. Due to this development, the total entropy on earth is increasing. This is a consequence of the 2nd law of thermodynamics.

Georgescu-Roegen described the fact that in modern times energy and matter are ultimately converted into waste in economic processes as follows: "The economic process is entropic: it neither creates nor consumes matter or energy, but only transforms low into high entropy" (Georgescu-Roegen, 1971, p. 281). He criticized that conventional economists do not take into account the 2nd law of thermodynamics, the law of entropy, although it affects the most important factor of production, the available (useful) energy. Therefore, as mentioned above, he considers the law of entropy to be "the most economical of all natural laws" (Georgescu-Roegen, 1971, p. 280). His statement: "Low entropy is a necessary condition for usefulness" (Georgescu-Roegen, 1979, p. 1042, our translation), also shows how highly he values the importance of entropy for the economy.

From a thermodynamic point of view, we can describe these processes as follows: The human economy is an open subsystem embedded in the larger system of the natural environment, the earth. The earth is to be considered a closed system, as it exchanges (almost) no matter with its environment, but energy, which enters in the form of solar radiation. Our economy still largely extracts materials with low entropy (oil, coal, natural gas, uranium, etc.) from the environment for the production of large quantities of capital and consumer goods. The means of production and ultimately also the consumer goods are mostly released back into the environment in the form of high entropy, i.e., a low state of order, after their use. Since the quantities of means of production have increased during industrialization, the carrying capacities of the environment are no longer large enough to process harmful joint products and reintegrate them into natural processes. As a result, our natural basis of life is, partly irreversibly, destroyed, as shown by species extinction. The task of environmental policy is therefore to stop and reverse this development. This can basically be done in two ways, which can be combined:

1. We can produce and consume less, thereby reducing the flow of energy and materials through our economy and at the same time the amount of pollutants that are released from the economy to the environment.
2. We can restructure our economy to a certain extent, particularly through a technological change, so that fewer pollutants are produced by using renewable energies and transforming into a circular economy. The use of renewable energies can significantly reduce the entropy caused by our economic activities; it is the basis for the transformation into a circular economy. Remaining amounts of pollutants must be so small that they can be disposed of in an environmentally friendly manner.

These paths have different chances of implementation. The first is opposed by the fact that many people still live today under a level of consumption that meets their basic needs and therefore needs to be raised. In addition, many people with a high level of material prosperity resist its reduction. In addition, we must still expect an increase in people. The second path is currently being pursued to a certain extent in parts of the world, but the developments are still far from an acceptable result, as shown by the increasing extinction of species, and still high amounts of greenhouse gas emissions per capita.

In addition, the thermodynamic perspective shows us the still unresolved challenges in the development of a circular economy. Because beyond the establishment of a supply system with renewable energies (which itself is associated with entropy), a large-scale implementation of recycling processes for particularly complexly connected materials, such as metals, rare earths, etc. (whose extraction, processing and recycling are all, to varying degrees, associated with entropy generation) is currently not foreseeable. This illustrates to us how the thermodynamic perspective can show us the importance of the first path, of producing and consuming less.

Important to Know: Significance of Thermodynamics for Research and Teaching

Since thermodynamics is crucial for all interactions between economy and environment, understanding it should be the basis of sustainable action. To this end, it is necessary to integrate central thermodynamic insights into teaching and research as well as in environmental and economic policy.

References[11]

Baumgärtner, S., Faber, M., & Schiller, J. (2006). *Joint production and responsibility in ecological economics: On the foundations of environmental policy*. Elgar.

BMVU. (2021). *Effizienz und Kosten: Lohnt sich der Betrieb eines Elektroautos?* https://www.bmuv.de/themen/luft-laerm-mobilitaet/verkehr/elektromobilitaet/effizienz-und-kosten

Faber, M., Manstetten, R., & Proops, J. (1996). *Ecological economics: Concepts and methods*. Elgar.

Faber, M., Niemes, H., & Stephan, G. (1983). *Entropie, . Umweltschutz und Rohstoffverbrauch: Eine naturwissenschaftlich ökonomische Untersuchung*. Springer. [Entropy, Environment and Resources. An Essay in Physico-Economics. *English Translation published 1987*].

Faber, M., Proops, J. L. R., & Baumgärtner, S. (1998). All production is joint production: A thermodynamic analysis. In S. Faucheux, J. Gowdy, & I. Nicolai (eds.), *Sustainability and firms, technological change and the regulatory environment, Edward Elgar, Cheltenham* (131–158). Elgar.

Georgescu-Roegen, N. (1971). *The entropy law and the economic process*. Harvard University Press.

Georgescu-Roegen, N. (1979). Energy analysis and economic valuation. *Southern Economic Journal, 45,* 1023–1058.

Hüfner, J., & Löhken, R. (2010). *Physik ohne Ende... Eine geführte Tour von Kopernikus bis Hawking*. Wiley-VCH.

Lauth, J. G., & Kowalczyk, J. (2015). *Thermodynamik*. Springer.

Tractebel Engie. (2018, Mai 18). *Rekordwirkungsgrad – 62,4% für ein Gas- und Dampf-Kombikraftwerk*. https://tractebel-engie.de/de/nachrichten/2018/rekordwirkungsgrad-fuer-ein-gud-kraftwerk.

[11]The contents of this concept are based on: Faber, M., Frick, M., Zahrnt, D. (2019) MINE Website, Thermodynamics, www.nature-economy.com.

Irreversibility: Why an Omelette can Never Become an Egg Again and What this has to do with Species Extinction

10

Contents

What's it About?

This is about irreversibility. When dealing with environmental problems, it is important to consider that certain developments cannot be undone, or can only be undone with enormous effort. Just as we intuitively understand that we cannot turn back time, we should also be aware that this is true for many processes concerning the environment.

Our efforts to preserve our natural life support systems will be more successful if we take into account the irreversibility of environmental damage, such as species extinction. Awareness of irreversibility is often not sufficiently developed in politics and economics. In economics, whose representatives still have a significant influence on politics and society, irreversibility has long played no or only a subordinate role. This continues to lead to significant weaknesses in dealing with environmental problems.

M. Faber et al., *Sustainable Action in Economy and Society*,
https://doi.org/10.1007/978-3-662-69122-9_10

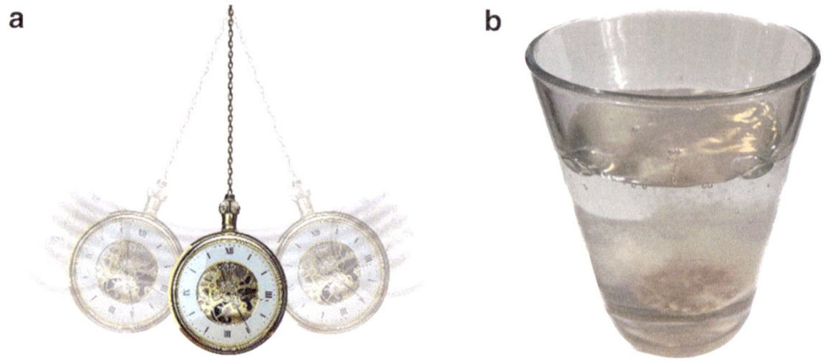

Fig. 10.1 **a)** Movement of a pendulum. (Source: Gerd Altmann, Pixabay.com). **b)** Effervescent tablet in a glass of water. (Source: Own illustration)

10.1 Introduction to the Concept

If we consider the movement of a pendulum (Fig. 10.1a), it is difficult to discern with the naked eye where the beginning and end of the movement lie. Imagine we were to film the pendulum's movement and play it backwards; it would not seem unnatural to us. This is because the pendulum's movement can be well described by the laws of Newtonian mechanics, which formed the basis of physics until the discovery of the laws of thermodynamics around 1850, as well as the theory of relativity and the discovery of quantum physics at the beginning of the 20th century. In the absence of energy losses due to friction or deformations—as is the case with the pendulum at least in a rough approximation—processes according to Newtonian mechanics are reversible in time. A process that would run backwards would be consistent with Newton's laws of gravitation and motion.

Now let's not imagine a pendulum, but an effervescent tablet in a glass of water (Fig. 10.1b) and also film that.[1] The contents of the effervescent tablet spread throughout the water in a very short time and are even partially released into the surrounding air. We intuitively understand that we cannot reverse this process; the film would seem unrealistic to us if it were running backwards. This is because entropy is created in this process, which roughly translated means that the process goes from an ordered to a disordered state (see Chap. 9 *Thermodynamics*). In theory, it is possible to reverse the process and return to the initial state, but only by supplying energy. In the case of the effervescent tablet, for example, it would be conceivable to hermetically seal the glass in the laboratory to prevent

[1] Both examples are inspired by the Science Slam "Entropy—From Cooling Towers and the Irreversibility of Things" presented by Martin Buchholz in the final of the first German Science Slam Championship. The video is available on Youtube (https://www.youtube.com/watch?v=z64PJwXy--8) (last accessed on 18.06.2023).

substances from escaping, distill the water, and reassemble the distributed components of the tablet. In the pendulum example, entropy is also created—due to friction with the air, etc.—but only very little, which is why the process is easily reversible with little effort.

Important to Know: Reversibility and Irreversibility of Processes
The irreversibility of processes is a consequence of the creation of entropy. According to the 2nd law of thermodynamics, the entropy created in a process is always greater, at best equal to zero, but never negative. The less entropy created (e.g., in a pendulum), the easier it is to reverse a process. The more entropy is created, the harder it is to reverse a process (e.g., effervescent tablet in a glass of water).

We conclude: While it is often easy to transform an ordered state into a state of disorder, the opposite direction is much more difficult, involves considerably more energy expenditure, and is sometimes even impossible. Unfortunately, the latter is exactly the task we are often confronted with when solving environmental problems. Through the mainly fossil-fueled high energy intensity of our economic activities, the global use and distribution of non-natural and partially toxic materials and substances, the destruction of the habitat of endangered species and other activities, we humans set many processes in motion that are not or only difficult to reverse. Our interventions in the environment are processes that cause very high entropy and we do not have sufficient clean energy sources to contain or reverse the chaos once caused. In short: We create more disorder than we can clean up (see Chap. 13 *Joint Production*).

In addition to the irreversibility caused by the creation of entropy, there is another dimension of irreversibility: Time. Some processes may be reversible in thermodynamic terms, for example, by supplying sufficient energy, but they are not in terms of time. We will examine this second, practical dimension of irreversibility in detail later in the chapter using the example of hydrogen production technologies.[2]

[2] As has been shown, thermodynamics provides a conceptual framework for dealing with temporal irreversibility: The concept of entropy is crucial for this framework. Further insights that can be gained for irreversible processes are presented in Faber, Manstetten & Proops (1996, pp. 103–112).

10.2 Reversibility and Irreversibility in Economics

The fact that the irreversibility of processes is often not taken into account is a phenomenon that also runs through the history of economics and has serious consequences for our handling of environmental problems. Leading economists of the 19th century, such as William Stanley Jevons (1835–1882) and Léon Walras (1834–1910), oriented themselves on the Newtonian conception of physics and time, which did not include thermodynamic laws. This led to the theory of general equilibrium, which forms the theoretical core of mainstream economics, being based on the reversibility of time. Although the equilibrium theory has since been extensively further developed in many respects, the problem of the conception of time largely remains. Time is often still viewed too statically and irreversible dynamic developments are not sufficiently taken into account (Faber et al., 1998, pp. 74–78). Within this static view, apples can be compared with pears and even today's apples with tomorrow's pears (Romer, 1994, p. 11). However, dynamic innovations (see Chap. 11 *Evolution*), such as the breeding of a new fruit variety, are not given the attention they deserve. As long as it's only about fruit, this may not be so problematic, but it becomes problematic when serious environmental problems need to be investigated and solved. Relying on an economic theory for environmental problems that delivers clear, quantifiable results, but which is based on unrealistic assumptions such as the static view and the reversibility of time and processes, can be seriously problematic for our dealing with environmental issues.

Building an economic theory that includes dynamic developments is extremely complex. Such an approach is pursued, for example, by the so-called *New Austrian Capital Theory,* which takes into account evolutionary developments (Faber et al., 1999; Stephan, 1995). Both the emergence of new things—such as climate-friendly methods of energy production –, the cost caused by maintaining the old—for example, the use of fossil fuels –, as well as the partly irreversible loss of natural resources and intact ecosystems, must be represented in a dynamic approach. Such an economic theory would be based on much more complex assumptions and it would not, or only very difficultly, be possible to deliver as clear and mathematically sound results as mainstream economics often can. However, the question must be asked whether it is better to base decisions on environmental issues on unrealistic assumptions, but clear results, or on realistic assumptions, but less clearly mathematically representable results. We do not want to give an answer to this question here, as both approaches have their respective advantages and disadvantages. However, we do point out that due to the irreversibility of many environmental problems, special caution should be exercised when decisions are made on the basis of simplified or even unrealistic assumptions.

10.3 Two Arrows of Time

The problem of irreversibility over time has been studied in the natural sciences in two different ways; these deal with two opposing aspects of the natural world. The first has already been discussed above, it is a tendency towards ever more mixing, as the example of the dissolving effervescent tablet in water shows. We never observe that pure water and an effervescent tablet are formed again from the mixture. All these phenomena are due to the Second Law of Thermodynamics. As explained in Chap. 9 *Thermodynamics*, these processes are associated with an increase in entropy. In an isolated system, this development can never be reversed. This circumstance has often been interpreted in the sense that the world is developing in one direction and is therefore asymmetric: from order to disorder. This view is captured under the term of the *First Arrow of Time*.

However, our experience with time is not limited to this negative view that everything is becoming more disordered; because in nature we observe development, structuring, growth, and self-organization. For example, we observe on the one hand how an apple tree decays at the end of its lifetime and thus behaves according to the *First Arrow of Time*. But what we mainly observe is the growth of the tree, the formation of branches, leaves, and apples. On a larger scale, we recognize how evolutionarily new plants and animals develop with their own self-organized life cycles. All of this has led to a diverse biodiversity over millions of years. Similarly, we see in the economy due to division of labor and technological progress a continuous evolutionary development, starting from hunter-gatherer societies, agricultural societies to today's industrial society. All these evolutionary tendencies towards self-organization (see Faber et al., 1996, pp. 110–111) are captured with the term of the *Second Arrow of Time*.[3] The driver of the *Second Arrow of Time* is energy, which is supplied to the system under consideration, for example, the Earth, from the outside. Phenomena of both arrows of time explain thermodynamically in very different ways irreversible processes. As mentioned above, ecological economic development can also be described and explained with this.[4]

[3] It was primarily the Russian-Belgian physicist, philosopher, and Nobel laureate Ilya Prigogine (1917–2003) who, together with his colleagues, developed a nonlinear thermodynamics to describe irreversible processes that are far from their equilibrium state, and used the term arrow of time for this.

[4] More detailed presentations of the two arrows of time can be found in Faber & Proops (1998, pp. 85–87) and in Faber, Manstetten & Proops (1996, pp. 103–114).

10.4 Irreversibility of Economic Policy Decisions: Technological Lock-In Effects

We have outlined above that the irreversibility of many processes is primarily based on thermodynamics. However, the irreversibility of crucial decisions in our economy, such as the development and spread of the internal combustion engine in the last century, is not solely physical. As Arthur (1989) explains, a particular technology does not necessarily prevail because of its technical superiority (e.g., higher efficiency), but that small, historical events can have a decisive influence. Although the steam engine initially had a higher potential economic efficiency, the internal combustion engine prevailed in cars. From the onset of significant economic scale effects in the production and distribution of a technology—that is, when production facilities are created on a large scale to mass-produce the product and thus lower the price—the switch to a possibly more efficient technology is made difficult for economic reasons, or even completely prevented. Once established production facilities and other infrastructure then cause a so-called *lock-in* or path dependencies. There is already too much capital tied up in the chosen technology for a switch to an alternative, even if better, technology to be worthwhile. The decision for the chosen technology is thus, for the time being, irreversible, despite a possibly socially undesirable result.

10.4.1 Hydrogen: What Color Should it Be?

We would like to illustrate the danger of such a lock-in, the irreversible choice of a technology path, using a currently open and frequently discussed question in the context of the energy transition to climate neutrality: Should blue hydrogen be used to decarbonize our economy?

First, some brief explanations on the question. Hydrogen is colorless in itself. The designation of hydrogen as blue, green, gray, yellow, or otherwise refers to the method of hydrogen production. We limit our consideration to the three production methods currently most used, or most discussed in the context of the energy transition. These are:

- Gray hydrogen: Production using fossil fuels (mostly natural gas).
- Blue hydrogen: Production using fossil fuels combined with *Carbon Capture and Storage* (CCS), i.e., the capture and storage of the greenhouse gases (GHGs) emitted during production.
- Green hydrogen: Production by water electrolysis using renewably generated electricity as an energy source.

Hydrogen will play a crucial role as an energy carrier in the global energy transition, as the use of hydrogen for transportation, production processes, electricity storage, etc. only produces water and oxygen as joint products and no climate-damaging GHGs (see Chap. 13 *Joint Production*). Although no GHGs are released when hydrogen is used, the situation is different during its production. Currently, the vast majority of the world's used hydrogen is gray (Metz et al., 2022), meaning that significant amounts of GHG are emitted by production using fossil fuels—mostly natural gas, but also coal. The demand for hydrogen will multiply in the coming decades in Germany and the whole world, and it must be ensured that these quantities are produced with as little GHG emissions as possible (German Energy Agency [dena], 2021; International Energy Agency [IEA], 2022; Ram et al., 2020). Green hydrogen, which only emits GHGs through the construction of renewable energy generation facilities (mostly wind and solar) and the production plant, the electrolyzer, is predestined for this as causes has low emissions.

However, green hydrogen is currently scarce, as it is very difficult in the short term to build up the necessary additional capacities for renewable electricity generation and electrolysis capacity (see Chap. 14 *Scarcity*). Different groups in politics, science, and industry are therefore calling for the construction of production facilities for blue hydrogen. The approach is as follows: To supplement existing technologies for the production of hydrogen based on the use of fossil raw materials such as oil or natural gas with CO_2 capture technology, thus bridging the short and medium-term scarcity of green hydrogen. The CO_2 released by the use of fossil raw materials would be separated from the exhaust air by a kind of filter and either stored in some form or used elsewhere. The problem: It is not guaranteed that the residual emissions of the production of blue hydrogen remaining after the CO_2 capture are really as low as is often assumed in these demands; recent studies have found that the GHG emissions from blue hydrogen could be just below the emissions from gray hydrogen (Howarth & Jacobson, 2021). For example, in the past, the climate-damaging methane leaks during the extraction and transport of natural gas have long been underestimated. Released methane has up to 30 times the greenhouse gas effect of CO_2. In addition, there is little experience with the large-scale use of CO_2 capture and subsequent storage.

This problem would not be so severe if we could easily switch to a "greener" technology should the fears about the climate-damaging effects of blue hydrogen be confirmed. However, this is where the importance of irreversibility becomes clear: There is a risk of a *lock-in* into fossil fuels and the technologies based on them. Extensive investments are necessary for the ramp-up of blue hydrogen production, i.e., the construction of new hydrogen production plants based on fossil energy carriers including CO_2 capture technology. This capital would be tied up in fossil technology for decades and would not be available for other technologies, such as wind turbines and photovoltaic systems, as well as plants for the production of green hydrogen, even if the production of blue hydrogen proves to be unsuitable. So there is a risk that the short-term incentive to invest in a less promising technology (at least from an environmental perspective) will lead to its implementation, bind investments, and hinder the necessary ramp-up of green hydrogen production.

10.5 Irreversibility—So What?

At this point in the chapter, one might ask oneself as an individual: What does the knowledge I have gained about irreversibility do for me? Of course, I cannot turn an omelette back into an egg, and it's understandable that certain directional decisions in the economy cannot be reversed, but how does this help me to better respond to environmental issues that are confronting me?

In response to these and similar questions, we will mention three characteristics of situations in which the concept of irreversibility is relevant.

1. In the Case of Possible Unwanted Outcomes of Actions

First of all, it can be said: If everything goes as desired, i.e. the outcome of our actions is positive, even with regard to possible consequences in the future, then irreversibility is not a problem at all. If we cannot undo something we do not want to undo, then that is not problematic. For example, if blue hydrogen proves to be the best technology with which we can achieve the energy transition, then it is not problematic if we cannot reverse the technology decision made. Irreversibility only becomes problematic in the case where we would like to reverse certain results of our actions, but cannot do so for technical or economic reasons. This is not always the case with environmental problems, but it is often the case.

2. Before the Actions

Furthermore, thinking about irreversibility is of little help when a problem is already present. Including irreversibility in environmental considerations after, for example, the production facilities for blue hydrogen have already been built, is not helpful. It is important to consider irreversibility in environmental issues before we act, i.e. *ex ante,* We should think today about which of our future actions in the environmental field are not, or only difficult to reverse, and let this knowledge flow into our decisions. The factor of time thus plays an important role in dealing with irreversibility. In Chap. 8 *Three Concepts of Time*, Kairos, the opportune time for an action, was introduced as a dimension of time. In relation to irreversibility, there is also a Kairos, namely the opportune time for thinking about irreversibility. This is usually before the start of the actual process, but at least before the last possible point of process termination.

3. In the Case of Ignorance

If we knew the outcome of our actions in advance, we would not need to deal with the irreversibility of these actions. We would only choose actions whose results would be desirable and which we would therefore not want to reverse. However, especially with actions that affect our environment, we have to deal with a lot of ignorance (see Chap. 12 *Ignorance*). Often we simply do not know the outcome of our actions, or we cannot foresee unwanted side effects in advance. Therefore, especially in connection with a lot of ignorance, it is necessary to weigh up whether possible unwanted consequences of an action are irreversible. In a further step, it can then be decided whether the hoped-for positive results are worth carrying out a possibly irreversible action.

10.6 Understanding Irreversibility as a Basis for Informed Decisions

The concept of irreversibility provides an important decision criterion for actions related to environmental problems. When it comes to solving environmental problems, or even preventing them from arising in the first place, we cannot avoid making decisions under risk, uncertainty, or ignorance (see Chap. 8 *Three Concepts of Time* and Chap. 12 *Ignorance*). Particularly serious is the fact that we can often reduce our ignorance, but never completely eliminate it. Therefore, there is always the possibility that actions will have undesirable consequences.

Reflecting on irreversibility helps us make better decisions, as we consider that some undesirable outcomes may be irreversible. This does not mean that we no longer carry out actions in the face of risk, uncertainties, or ignorance. If this were the case, caution would paralyze us and important actions with positive results would no longer be taken. Therefore, it is important to expand our decision-making basis by reflecting on irreversibility and thus make better decisions based on a broader information base. Which decisions are made and whether irreversibility is decisive for this decision is up to the individual and the judgement used (see Chap. 7).

In relation to the initial question about blue hydrogen, this means that a politician or a businessman, for example, who want to answer the question for the orientation of their politics or his company, would not necessarily decide against blue hydrogen just because the result could have irreversible consequences. However, the possible irreversibility of such a decision would provide information that could help the politician or businessman make a better decision.

Important to Know: The Practical Significance of Irreversibility

- **Irreversibility due to Lock-In:** Irreversibility can be both physical (see explanations on entropy in Sect. 10.1 above), as well as economically justified, when certain technologies tie up the necessary capital, making it no longer available for alternatives. A case in which it is no longer possible to switch from a suboptimal technology to a better one due to previous decisions (investments, laws, infrastructures, etc.) is referred to as *Lock-In*.
- **When does Irreversibility Play a Role in Our Decisions?** Reflection on irreversibility is necessary when
  - Results of actions may be undesirable.
  - Actions are still pending and results are still influenceable.
  - When there is ignorance about the outcome of actions.

- **Irreversibility as an Extension of the Decision-Making Basis for solving Environmental Issues:** Considerations about the irreversibility of actions and their results do not solely determine decision-making, but expand the information base on which more informed and possibly better decisions can be made.

References[5]

Arthur, W. B. (1989). Competing technologies, increasing returns, and lock-in by historical events. *The Economic Journal, 99*(394), 116–131.

Deutsche Energie-Agentur GmbH (ed.) (dena). (2021). d*ena-Leitstudie Aufbruch Klimaneutralität.*

Faber, M., Manstetten, R., & Proops, J. L. R. (1996). *Ecological economics: Concepts and methods*. Elgar.

Faber, M., & Proops, J. L. R. (1998). *Evolution, time, production and the environment*. Springer Science & Business Media.

Faber, M., Proops, J. L. R., & Baumgärtner, S. (1998). All production is joint production: A thermodynamic analysis. In S. Faucheux, J. Gowdy, & I. Nicolai (eds.), *Sustainability and Firms, Technological Change and the Regulatory Environment* (131–158). Elgar.

Faber, M., Proops, J., Speck, S., & Jöst, F. (1999). *Capital and time in ecological economics: Neo-Austrian modelling*. Elgar.

Howarth, R. W., & Jacobson, M. Z. (2021). How green is blue hydrogen? *Energy Science & Engineering, 9*(10), 1676–1687.

International Energy Agency (ed.) (IEA). (2022). *Net Zero by 2050: A Roadmap for the Global Energy Sector.*

Metz, S., Smolinka, T., Bernäcker, C. I., Loos, S., Rauscher, T., Röntzsch, L., Arnold, M., Görne, A. L., Jahn, M., & Kusnezoff, M. (2022). Wasserstofferzeugung durch Elektrolyse und weitere Verfahren. In *Wasserstofftechnologien* (207–258). Springer.

Ram, M., Galimova, T., Bogdanov, D., Fasihi, M., Gulagi, A., Breyer, C., Micheli, M., & Crone, K. (2020). *Powerfuels in a renewable energy world-global volumes, costs, and trading 2030 to 2050*. LUT University und Deutsche Energie-Agentur GmbH.

Romer, P. (1994). New goods, old theory, and the welfare costs of trade restrictions. *Journal of Development Economics, 43*(1), 5–38.

Stephan, G. (1995). *Introduction into capital theory: A Neo-Austrian perspective*. Springer Science & Business Media.

[5]The contents of this concept are based on: Faber, M., Frick, M., Zahrnt, D. (2019) MINE Website, Irreversibility, www.nature-economy.com

11 Evolution: Everything Evolves—Possibilities and Limits of Our View Into the Future

Contents

What's it About?

This is about evolution. How will the economy evolve in the future? Is it possible to halt the ongoing environmental destruction, or do the existing structures provide too little scope for change? Can we actively influence these changes? If so, how? We attempt to answer these and similar questions with the concept of evolution.

This chapter demonstrates the potential of the evolution concept, which takes into account both predictable and unpredictable processes, inventions, innovations, novelty, and ignorance. This understanding aids in promoting sustainable development, as we can influence and potentially control developments more specifically with the knowledge of evolutionary processes.

M. Faber et al., *Sustainable Action in Economy and Society*,
https://doi.org/10.1007/978-3-662-69122-9_11

11.1 Introduction to the Concept[1]

The concept of evolution was originally developed by the biologist and naturalist Charles Darwin (1809–1882), who used it to explain the development of species. He used the terms phenotype and genotype, as they are commonly used in biology today.

Consider an organism: it has a certain appearance, characteristics, and abilities. All of these constitute its phenotype, which is determined by two factors. One is the potential that the organism has inherited from its parents, its genetic structure (DNA), or its genotype. The second factor is its environment. For example, if we start with two organisms with the same genotype, as is the case with identical twins, they will grow differently and have different abilities, even with their nearly identical DNA, if they are fed or raised differently.

It is generally accepted that in biological systems the genotype influences the phenotype, but the phenotype does not influence the genotype. A spontaneously occurring or artificially induced, permanent change in the genetic material of a cell causes a mutation of the genotype. This can, for example, cause the arms of a monkey to become stronger. However, if a monkey strengthens its arms, its offspring will not inherit this characteristic.

These considerations show that the genotype of an organism describes the possibilities or potential of this organism. A corresponding phenotype represents *one* of these possibilities. A genotype thus allows for different phenotypes. Which one is realised depends on the environmental conditions.

Important to Know: Genotype and Phenotype
A natural organism has a genotype and a phenotype.
The genotype is influenced by its genetic structure (its DNA).
The appearance of a natural organism is its phenotype. It is determined by its gene structure and its environmental conditions. The genotype determines the phenotype alongside the environmental conditions, but the phenotype does not determine the genotype.

The concept of evolution based on the interplay of genotype and phenotype can be generalised so that it can also be used outside of biology for developments in other scientific areas. Particularly relevant for sustainability issues are the study of evolutionary development processes in the economy and their interaction with the environment.

[1] This chapter utilises material from the book by Malte Faber and John Proops (1998, Chap. 1–3).

The current understanding of evolution in the public is that conditions in a society improve over time. Examples of this are:

- Evolution of technology through technological progress.
- Evolution of the economy through economic growth.

This understanding is insufficient for questions about the future of humanity; because we must consider that nature is limited. This is particularly true for the availability of raw materials and the capacity to absorb pollutants in water, air, and soil.

Consequently, our task is to develop a comprehensive evolutionary approach that is suitable for ecological-economic investigations. It should enable interdisciplinary investigation of evolutionary processes in

- physical,
- biological,
- ecological, and
- economic

systems; because all these systems are important for our question of predictability and influence on ecological-economic relationships.

Before we proceed, we want to formulate the term evolution in such a way that it is suitable for the comprehensive task of predicting ecological-economic relationships. It is obvious that a broad definition of evolution is required for this. For our purposes, the key characteristics of evolution are:

- Change and
- Time.

These considerations lead to the following definition:

Evolution is the process of change of something over time.

Instead of an analytical definition, we have used a general one. This formulation is so general that it is suitable to be used in different contexts.

Important to Know: What Does Evolution Mean?
Definition: *Evolution is the process of change of something over time.*

This is a general and broad conceptualisation of evolution. It allows recognising and investigating evolution in different areas of reality and sciences.

11.2 Conceptual Evolutionary Framework

We aim to develop a general conceptual evolutionary framework to investigate ecological-economic relationships. We are interested in the potentialities and realizations of physical, biological, and economic systems over time. For simplicity, we use the term "phenotype" when referring to "realizations" and "genotype" when referring to "potentialities", based on our general conceptualization of the concept of evolution. We do not draw analogies to biology, but acknowledge that these terms were first used in biology and expand them for our broader considerations.

We have already discussed the interplay of genotype and phenotype in biology. Now, let's consider evolution in physical systems. How can we determine the *genotype of physics*? Processes in physical systems are determined by fundamental physical constants and laws of nature, as well as by environmental conditions. The realization of these possibilities is the observed behaviour of the respective physical system.

The genotype of a physical system, according to our definition, results from the possibilities, i.e., from the fundamental natural constants and laws of nature. Accordingly, the phenotype of a physical system is a realization of these possibilities in interaction with the respective environmental conditions. This conceptualization leads to the conclusion that, unlike in biology, all physical systems have the same genotype. Different physical phenotypes, therefore, do not originate from different genotypes. As in biology, a physical phenotype cannot influence its physical genotype.

As in biology and physics, we can characterize an economy in economics by its potentialities and realizations, i.e., by distinguishing between the genotype and phenotype of the economy. However, the determination of genotype and phenotype in the economy is not as clear-cut as it is in biology or physics. This is because both the genotype and the phenotype of the economy are composed of a multitude of factors that change rapidly over time. Moreover, certain manifestations of the phenotype can change the genotype. We will revisit these unique characteristics of genotype and phenotype in the economy later in this chapter.

First, we ask: What constitutes the genotype of an economy? To answer this, we need to consider which factors determine the potentialities of an economy. These can be grouped into factors that relate to the supply in the economy, i.e., the production of goods and services, and factors that affect the demand side.

The production of goods on the *supply side* is mainly influenced by

- the geographical conditions,
- the raw material endowment,
- the availability of capital goods (such as machines, tools, buildings, infrastructure, transport vehicles),
- the existing technology,
- the labour force and their training,

- the legal, economic, and social institutions, and
- the form of economy (be it market economy, socialism, or planned economy).

On the *demand side*, the actual production of goods in a market economy is significantly shaped by

- the consumption habits or preferences of economic actors,
- their income,
- their wealth, and
- the "worldview" of economic actors, including religious and social norms.

We explained earlier that the phenotype in biology results from the interplay between the genotype and the natural environmental conditions. The same applies to the phenotype of an economy. This implies that, even if the genotype of an economy does not change, the phenotype can change solely due to environmental changes, such as climate changes.

The *phenotype of an economy*—for example, a market economy—results in a given period mainly from the actual

- consumed quantities of raw materials,
- the technologies used,
- the types of capital goods used,
- the use of labour,
- the produced quantities of consumer goods,
- the investments in new capital goods,
- the prices resulting from supply and demand,
- the demanded quantities, as well as the distribution of consumer goods, income, and wealth among the economic actors.

We will see later that these differences between biology, physics, and economics result in considerable differences in the possibilities of predicting the course of evolutionary processes.

11.3 Evolutionary Differences Among Biology, Physics, and Economics

We aim to explain a significant evolutionary difference between physics and biology on one side, and economics on the other. Neither in physics nor in biology does the phenotype influence the genotype. The physical genotype has remained unchanged throughout the universe since the Big Bang 13.8 billion years ago. The genotype in biology does change, but not due to influences of the phenotype. Instead, it changes through random mutations of species—a process that usually occurs very slowly. In economics, however, the phenotype can bring about decisive changes in the genotype. In hunter-gatherer societies, these changes occurred slowly over several centuries or even millennia. Today, the genotype of the economy is changing rapidly due to the consumption of non-renewable resources, accumulation of pollutants in water, soil, and air, destruction of biodiversity, and especially technical progress and education, often faster than the economic phenotype.

Key Points: Evolutionary Processes from the Perspective of Different Disciplines

- Evolutionary processes in physics, biology, and economics can be studied using their genotypes and phenotypes. The genotypes describe the potentials or possibilities, while the phenotypes describe their realizations.
- The phenotypes in physics and biology cannot influence their genotypes. This is quite different in economics, where the phenotype can change the genotype.
- The physical genotype has not changed since the Big Bang. The biological genotype changes relatively slowly due to mutations. The genotype of the economy changed slowly in hunter-gatherer societies, but it is currently evolving very quickly.

11.4 Long-term Evolutionary Developments

Current discussions about environmental problems often emphasize their novelty, suggesting that in earlier times, humans lived in harmony with their environment and nature. However, the interaction between the economy and the environment is as old as humanity itself when viewed in the long term. There is ample evidence that even hunter-gatherer societies significantly altered their natural environment through their economic activities. For instance, after humans arrived on the American continent about 12,000 years ago, there was a massive extinction of many large mammals, and it is assumed that this was due to excessive hunting (Crosby, 1986). Similar experiences occurred in Australia after humans migrated there about 40,000 years ago. The first Australians systematically altered their ecosystem over several thousand years by using fire to improve their hunting opportunities (Blainey, 1983). Over several millennia, the genotype of the economy—and thus also the phenotype—was greatly altered by economic activities. While the evolutionary process in these examples occurred slowly, allowing humans to react to the changed conditions, the speed of evolutionary processes has accelerated so much with the ongoing development over the last three centuries that we, as humanity, struggle to deal with the drastic changes that come with it. We illustrate the problems of such accelerated evolutionary processes using the example of the development of industrial agriculture.

11.4.1 Rise of Agriculture and Industrialization

The rise of agriculture in the last millennia has led to a drastic and far-reaching transformation of the natural environment. In Europe, for example, people became accustomed to many mountain grasses and heath plants. However, this vegetation is not natural, but the result of deforestation by Neolithic[2] farmers and was maintained over thousands of years by the grazing of domesticated animals (Hoskins, 1970).

During the era of hunters and gatherers and agriculture, the natural environment was affected relatively slowly. The transition to agrarian societies, and especially the development of industrialization in the last four hundred years, dramatically changed the speed, extent, and nature of the interactions between the economy and the environment. In agrarian societies, the most important building material and main fuel was wood, meaning these economic forms were almost exclusively dependent on renewable natural resources. This changed with the onset of industrialization in England in the middle of the 18th century.

During industrialization, a comprehensive change in the genotype of the economy occurred with the introduction of the steam engine. As a result, the phenotype also changed: Much more wood was needed than in agrarian societies. As industrialization progressed, England's forests became increasingly scarce, creating a pressure for innovation. This led to further changes in the genotype: The production machines were converted to coal as an energy source, later also to oil and gas, and even later to nuclear energy. This development continues to impact us today, leading to the enrichment of the atmosphere with greenhouse gases and creating a new pressure for innovation: Coal-based fuels must be replaced by renewable energies to prevent further endangerment of the planet's habitability by humans.

Important to Know: Historical Experiences

From the consideration of historical developments, we can learn that changes in the environment lead to changes in the economy, as in biology, where not every genotype can assert itself. However, it is important to note that these changes come with consequences. In biology, it is the extinction of certain species, while in the economy, it is the necessity to partly abandon or adapt old forms of production and consumption and to develop new ones.

[2]The Neolithic or New Stone Age, technically Neolithic (from ancient Greek neos 'new, young' and lithos 'stone'), is an epoch of human history defined as the transition from hunter-gatherer cultures to shepherd and farmer cultures. It lasted from about 10,000 to 3,000 CE, i.e., until the beginning of metal processing.

11.5 The Triangle of Causation

The shift from an economy fundamentally based on renewable resources such as wood and natural resources, to an economy dependent on non-renewable fossil fuels, has largely influenced the dynamics of modern economic activity and many of the current environmental problems. Today, the pressure of environmental issues necessitates a fundamental transformation of the economy. These processes, described in the previous section, follow an evolutionary pattern known as the *Triangle of Causation,* which is illustrated in Fig. 11.1.

When an invention is made, such as the development of the steam engine, it represents a change in the economy's genotype. The potential to use this invention exists. If it is utilized—as was the case with the steam engine—it leads to innovation. This changes the mode of production in the economy and its effects on the environment, constituting a change in the economy's phenotype. However, the environmental effects do not mark the end of the process. In certain cases, they lead to further changes in the genotype. This occurs when the economy's production methods lead to a shortage of raw materials and/or environmental damage. For instance, the deforestation during the Industrial Revolution in England made it impossible to continue production methods dependent on wood as an energy source. The result was a new invention, the steam engine, which enabled the use of coal as an energy source—a further change in the genotype, which in turn spurred further changes in the phenotype.

Three points should be noted:

1. The process of traversing the "Triangle of Causation" is a temporal process that can span years, during which changes in the genotype and phenotype occur repeatedly.
2. There is no fixed beginning or end to the process.
3. Even a random environmental change—such as an earthquake or an epidemic—can initiate the process.

However, this representation of a single triangle of an economic sector is insufficient to describe historical technological developments. For instance, the need for coal led to the development of steam-powered pumps, which enabled deeper coal mining. This use of the steam engine subsequently led to changes in a completely different economic sector: the transport sector. Both on land and water, locomotives and ships could now be powered by the steam engine, which resulted in countless and often unpredictable further developments.

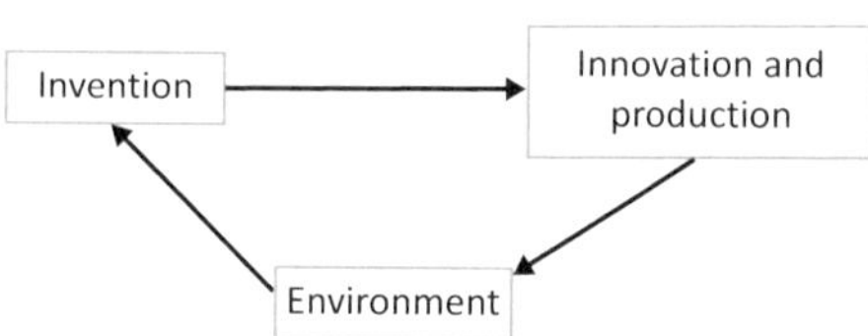

Fig. 11.1 The Triangle of Causation: Long-term economic and environmental interactions

Therefore, it is important to note the following about the *Triangle of Causation* from Fig. 11.1:

- The Triangles of Causation can come to an end over time if no further inventions are made. An example of this is the music box industry, which began in the Black Forest in 1796 and ended around 1930 because it could not compete with the gramophone.
- There are different Triangles of Causation with different phases within an industry, between industries, and between economies.
- Inventions in one triangle, as explained using the example of the steam engine, can trigger other Triangles of Causation, which in turn develop their own dynamics.

11.6 Insights from the Historical Interaction Between Economy and Environment

The complex, recursive relationship between economic development and the natural environment is far from simple. However, we can identify three significant elements.

1. The consumption of non-renewable natural resources is irreversible over time (see Chap. 10 *Irreversibility*). Any technology that relies on finite, non-renewable resources will eventually become obsolete.
2. The invention and subsequent innovation of new techniques have enabled the continuation of economic activities despite the depletion of natural resources. This can be achieved through more efficient utilization, i.e., through resource-saving inventions, or through the use of natural resources that previously had no value but have become useful through an invention. This is often referred to as resource substitution. It is clear that an invention and its subsequent innovation represent an irreversible process, as the invention introduces a novelty. Consequently, the genotype of the economy changes.
3. Each innovation cycle has a specific time structure (see Chap. 8 *Three Concepts of Time*). This begins with an invention, includes the gradual accumulation of capital goods necessary for the dissemination of the invention, and the reduction of the stock of capital goods of the previously used technology.

Important to Know: Predictability and Unpredictability

The historical development of the interaction between the economy and the natural environment is characterized by certain properties. These can be explained and *ex post* (in retrospect) represented on the basis of models. They can also be used to make predictions about the future by trying to model these *ex ante* (in advance). However, there are events, such as earthquakes, which can only be predicted very inaccurately, if at all, due to their nature. This class of events includes all those where novelty occurs, such as through inventions, revolutions, wars, or epidemics and ignorance (see Chap. 12 *Ignorance*).

11.7 Evolution and Environmental Problems: The Question of Predictability

The solutions for existing and upcoming environmental problems lie in the future. If one wants to contribute, it is necessary to first fundamentally deal with the question of what we can know about the future. Because long-term developments confront us with the problem of predictability (see Chap. 12 *Ignorance*) and especially the emergence of novelty. Changes in a genotype are difficult or even impossible to predict in advance, which shows the limits of our predictions. It is also challenging to predict how the phenotype will develop from the possibilities available in the genotype and the environmental conditions. How can we deal with all these difficulties of predictability?

11.7.1 Equilibrium and Evolution

Before discussing the relationship between predictability and evolutionary development, it is beneficial to consider the concept of equilibrium and its relation to our understanding of evolution. As we will demonstrate, equilibria provide guidance regarding the direction of evolutionary developments.

Firstly, let us define two terms: A system has a *static equilibrium* if there is a state A characterized by the following properties: If the system is in state A, it will remain in state A, provided it is not disturbed externally.

A system has a *dynamic equilibrium* if there is a sequence of consecutive states that always repeat at equal time intervals. For instance, a system consistently transitions over the course of T periods from state A → state B → state C → state A. Again, this assumes that the system is not externally disturbed.[3]

[3] A static equilibrium is thus a special case of a dynamic equilibrium, in which all states of the cycle are identical.

Key Concepts: Static and Dynamic Equilibrium

- **Static Equilibrium:** In a system, no change occurs or changes always lead to the same final state A. An example of a static equilibrium is a stone lying on the ground.
- **Dynamic Equilibrium:** Changes occur cyclically and according to the same pattern, so it is predictable that state A → state B and state B → state C follows, which in turn is replaced by state A. A dynamic equilibrium can contain any number of states. An example of a dynamic equilibrium is a pendulum.

We will now examine the relationships between equilibrium and evolution. Firstly, we acknowledge that an evolutionary development does not necessarily have an equilibrium. Conversely, if a system, such as an economy, is always in equilibrium, then it cannot evolve. We can also assert that evolutionary systems can have one or more equilibria, both static and dynamic. While the concept of equilibrium is highly abstract, evolution can be continuously observed in reality.

Different scientific fields exhibit varying degrees and forms of abstraction. We will illustrate this using examples. In physics, equilibria can be directly observed, for example, a stone on earth is in a static equilibrium, a freely swinging pendulum in a dynamic equilibrium. We can easily recognize both phenomena with our eyes. This is not the case in biology, as a biological system is much more challenging to observe due to its heterogeneity and the multiple connections and relationships within the system. Consequently, biological equilibria must be determined by meticulous measurements and calculations. For the same reason, determining an equilibrium in the economy is even more challenging. Particularly in the current economy, due to its dynamic nature, equilibria are often not reached at all, as changes such as technological progress, population growth, and changes in environmental conditions quickly alter the economy's genotype. We will discuss this further below.

We can already discern here that the reason for these differences in the determination of equilibria lies in the differences in the genotypes of these sciences, as previously mentioned.

Key Concepts: Genotypes and Equilibria in Different Sciences

In physics, the genotype does not change, in biology it changes slowly, and in the economy it changes rapidly. Generally, equilibria can only occur when the genotype does not change. A change in the genotype disrupts existing equilibria until a new equilibrium forms with a new genotype. However, if the genotype changes very quickly and repeatedly—as is the case in our current economy—there is rarely or no equilibrium.

11.7.2 Predictability and Unpredictability: The Relationship between Equilibria and Evolution

When we explore the question of what we can predict and what we cannot, a useful starting point is the consideration of the genotype. It is crucial to distinguish whether the genotype of the system we want to discuss is constant or not.

If the genotype is constant, equilibria can occur. These allow a certain degree of predictability, either because nothing changes *(static equilibrium),* or because certain changes occur cyclically according to the same pattern *(dynamic equilibrium).* In physics, such an equilibrium is, for example, the regular movement of planets. We can predict with great certainty where a planet will be at a specific point in the future. In biology, dynamic equilibria occur, for example, in the population of animals and plants, while static equilibria can be identified in the tree population of primeval forests. However, equilibria can be disturbed, even if the genotype remains constant. This happens when the system is altered by environmental influences. In biological systems, this is the case when the basic conditions of life for animals and plants change, for example due to environmental pollution or habitat reduction. Environmental conditions also influence economic systems, not only by exerting pressure for change on the genotype, but by causing new expressions of the phenotype. This also happens in the economy through social, cultural, and political influences. For example, Germany has temporarily reduced its natural gas consumption due to the effects of the Ukraine war and is sourcing remaining quantities in the winter of 2022/23 from non-Russian sources—a short-term change in the phenotype, but not in the genotype. In general, strong influences on a system under consideration lead to the dissolution of equilibrium states at least in the short term and therefore make prediction more difficult.

Similar to environmental influences, changes in the genotype also affect predictability. When the genotype of a system changes, the system is taken out of its equilibrium. Over time, if the new genotype remains constant, a new equilibrium can be established and predictions can be made. However, in modern economic systems, where the genotype changes very quickly, equilibrium states are rare. Currently, it is difficult to foresee in which direction areas such as artificial intelligence, genetic engineering, telecommunications, or transport and traffic will develop, as new inventions are constantly being made and many are or will be widely used. Under these circumstances, accurate predictions seem almost impossible.

This leads us to the question of whether we can predict economic developments at all, especially in connection with environmental problems. The answer is: Yes, we can! Although we can hardly foresee developments in various areas of the economy, these developments always represent only a part of the genotype of the economy. While the genotype in the field of artificial intelligence, for example, is changing rapidly, other areas of the genotype of the economy, such as power plants operated with fossil fuels, remain constant, at least over longer periods of time.

For example, the entire physical genotype, which is also part of the economy, never changes. Every invention, no matter how innovative, is subject to the laws of *thermodynamics* (see Chap. 9). For example, we can predict today that the efficiency of coal-fired power plants cannot be significantly increased any further, somewhere in the range around 50% no further improvement is possible. Similarly, an internal combustion engine will never achieve the efficiencies of modern electric motors. Another area of the genotype of the economy that has remained constant for a long time to this day is the genetic composition of humans. Even though people differ greatly, for example in their worldviews or consumption preferences, large areas of our lives have remained the same for thousands of years. We need to eat, sleep, drink enough, we are social beings, live in groups, engage in cultural activities, etc.

Physics and the genetic makeup of humans are just two of many other areas of the genotype of the economy that show a certain constancy. For predictions about the future, it is therefore important to recognize these areas, and, if possible, to identify resulting equilibria. In combination with knowledge about areas where the genotype changes very quickly or where unpredictable environmental influences regularly dissolve equilibria, these insights provide orientation in the question of what we can know about the future and what we cannot (see Chap. 12 *Ignorance*). This varies greatly: Sometimes we can know nothing, sometimes very little, sometimes something, sometimes a lot, or even a great deal. Depending on how quickly genotypes change and how strongly the environment influences the system under consideration from the outside, we have a chance to gain knowledge about the future through predictions. These assessments of the possibilities of our knowledge about the future then also allow us to see possible solutions to environmental problems from a different perspective.

Summary
In this chapter, we have developed a general conceptualization of evolution. We have defined evolution as the process of change of something over time. The starting point of our considerations was that biology, with its concepts of genotype (possibilities) and phenotype (realizations), provides crucial clues for this conceptualization. The former allow the emergence of novelty, while the latter do not. Both terms can be used for physics and economics in addition to biology. The evolutionary perspective shows us what and to what extent predictions about the future are possible. In this way, we can estimate the limits of our knowledge about future developments. This ability is of great importance for environmental policy decisions.

References[4]

Blainey, G. (1983). *Triumph of the nomads: A history of ancient Australia*. Pan Australia.
Crosby, A. W. (1986). *Ecological imperialism: Biological and cultural consequences of 1492*. Cambridge University Press.
Faber, M., & Proops, J. L. (1998). *Evolution, time, production and the environment*. Springer Science & Business Media.
Hoskins, W. G. (1970). *Making of the English landscape*. Hodder and Stoughton.

Reading Recommendations for Further Reading on Part 3 "Time and Nature"

Barbehön, M. (2023). *Zeichen der Zeit. Umrisse einer Politischen Theorie der Temporalität*. Verlag. [Three Concepts of Time]
Boulding, K. (1981). *Evolutionary economics*. Sage. [Evolution]
Georgescu-Roegen, N. (1971). *The entropy law and the economic process*. Harvard University Press. [Thermodynamics and Joint Production]
Hawking, S. (2009). *A brief history of time: From big bang to black holes*. Random House. [Three Concepts of Time]
Hüfner, J., & Löhken, R. (2010). *Physik ohne Ende… Eine geführte Tour von Kopernikus bis Hawking*. Wiley-VCH. [Thermodynamics]
Prigogine, I. (1980). *From Being to Becoming Time and Complexity in the Physical Sciences*. Freeman. [Irreversibility]
Proops, J. (1985). Thermodynamics and economy: From analogy to physical functioning. In W. van Gool & J. J. C. Brugginck (eds.), *Energy and time in the economic and physical sciences* (pp. 155–174). North-Holland Publishing Company. [Thermodynamics]
Schiller, J. (2002). *Umweltprobleme und Zeit: Bestände als konzeptionelle Grundlage ökologischer Ökonomik*. Metropolis. [Three Concepts of Time and Stocks]
Schumpeter, J. (1911). *Theorie der wirtschaftlichen Entwicklung*. Duncker und Humblot. [Evolution]

[4]The contents of this concept are based on: Faber, M., Frick, M., Zahrnt, D. (2019) MINE Website, Evolution, www.nature-economy.com.

Part IV
The Interplay of Humans and Nature

Ignorance: How Knowledge About our Ignorance Helps with Environmental Problems

12

Contents

What's it About?

This text explores the concepts of knowledge and ignorance. It questions how we define the boundaries of our knowledge and to what extent ignorance can be reduced. It asks what humans can know and what systematically eludes their understanding.

The text also examines the extent to which humanity overestimates its knowledge and underestimates its ignorance, for instance, in relation to technical solutions or spontaneous social developments. It considers the implications of this for the controllability of technology, nature, and society, and how people can productively deal with their ignorance in relation to environmental problems.

M. Faber et al., *Sustainable Action in Economy and Society*,
https://doi.org/10.1007/978-3-662-69122-9_12

12.1 Introduction to the Concept

The topic of this chapter may seem unusual for a scientific book at first glance. When we discuss *ignorance*, we are dealing with something that we generally assume is reduced or even eliminated by scientific research. It is therefore surprising to learn that both the importance of ignorance and the knowledge about it in our more than 40 years of thinking and researching societal and ecological problems has increased rather than decreased.

This increase is related to our specific perspective, but also to the fact that due to the successes of science, research, education, and entirely new forms of knowledge transfer (Wikipedia, Youtube, Social Media, E-Learning), societal focus is often directed towards what we already know, think we know, or should know. In doing so, it is overlooked how important it is for finding good solutions and decisions to always consider that there are facts, relationships, and events about which we systematically know nothing and often cannot know anything.

To put it concretely, without realizing it, we often find ourselves in situations where we systematically *overestimate* our *knowledge* and systematically *underestimate* our *ignorance*.

Even our everyday actions are shaped more than we realize by the fact that we are ignorant in many regards. Accordingly, we should pay much more attention to this fact. But let us take one thing at a time.

Firstly, we want to explain how we come to the assertion that ignorance plays a much larger role (especially in dealing with environmental problems) than we generally assume. We have pointed out at different places in this book that environmental problems usually arise from a combination of natural, societal, and economic factors over a longer period of time (see Chap. 8 *Three Dimensions of Time,* Chap. 11 *Evolution,* Chap. 13 *Joint Production*). It is difficult to monitor the functioning of the various sub-systems of society, such as the economy, politics and civil society. It becomes even more complicated when we add interactions between these societal subsystems and natural ecosystems—for example, considering that we use natural resources for many economic activities and return production waste as emissions or garbage to nature. Taking into account that all these interactions happen over a long period of time and the effects sometimes only show up with a delay, the search for solutions becomes a task that literally gives us a headache.

Based on this schematic representation, it becomes easily recognizable that due to the complex interactions that structurally underlie environmental problems, we initially do not—and often never—know all the facts and understand all the relationships. So we cannot know everything we would need to know to understand the problem. And it is also understandable that long periods of time bring with them a multitude of changes that we cannot know or predict in advance.

12.1.1 Four Forms of Ignorance and How They Influence Our Ability to Act

This applies to us as individuals, who quickly reach the limits of our knowledge simply because we cannot possibly be experts in all areas of knowledge that may be relevant. We are talking here about *individual ignorance.* But it also applies to us as societies, because certain facts and events cannot be known or predicted even with the best science and research. In this case, we speak of *collective ignorance.*

In both cases, with *individual* as well as *collective ignorance,* we must still distinguish further. Because even if the described interactions are complex and long periods of time are fraught with uncertainties, we can still find out quite a bit with some time and effort. By researching, learning and further educating ourselves as individuals, or by promoting research and initiating a scientific discussion on certain topics as a society, we can gain knowledge, *reduce* our ignorance and even overcome it in some places. In this way, if we have the necessary time and the appropriate resources available, we can make quite a bit of progress.

However, even with infinite time and resources, we cannot completely overcome ignorance in all areas and questions. This is because, in addition to *reducible ignorance*, there is always an area of *non-reducible ignorance*. There can be various reasons for this, extending into complex epistemological questions. However, we do not need to delve that deep at this point. It suffices to acknowledge that there are instances where events occurred that were practically unpredictable. In such cases, we refer to these as *novelties,* which we simply could not have anticipated. We will provide three examples:

The first example is the astonishing and fundamental change that Greta Thunberg brought about in the global climate policy debate with her initially solitary school strike for the climate. No one could have predicted that just a few months after she first sat in front of the Swedish Parliament with her homemade sign, millions of people worldwide would take to the streets for more climate protection under her slogan "Fridays for Future". Even seasoned experts on social systems and their interactions could not have predicted that a 15-year-old Swedish schoolgirl would deliver the most sensational speech at the meeting of the world's most powerful people at the World Economic Forum in Davos just a few months after her individual decision to strike. Yet, Thunberg's wake-up call "Our house is on fire" is today one of the most defining slogans and unites many people who are committed to more climate protection.

The second example of *novelty* is how the earthquake off the Japanese coast in 2011 and the resulting nuclear accident at the Fukushima nuclear power plant led to a turnaround in energy policy by the German federal government within just one week. Just a few months after she had extended the running times of German nuclear power plants with full conviction, Angela Merkel's government now suddenly and without hesitation declared a complete nuclear phase-out by 2022. No political analyst, no matter how experienced, no interdisciplinary research group on energy policy, no matter how high-ranking, would have been able to predict this paradigm shift.

The third example is the widespread use of chlorofluorocarbons, so-called CFCs, in the 20th century. This chemical compound was indispensable from people's everyday lives for a long time, as it has many advantageous properties: it is not toxic, not flammable, and can be used for very different purposes, especially for thermal insulation and cooling. It is therefore not surprising that CFCs were quickly produced and used in large quantities after mass production was made possible in the 1930s. They provided practical services in refrigerators, freezers, refrigerated trucks, or in large cold storage houses and served as propellants in aerosol cans as well as cleaning and solvent agents. The fact that the use of the practical all-rounders CFCs had a very dangerous downside and their widespread use posed an existential risk to people only became clear when great damage had already been done: In 1985 it was discovered that CFCs rise into the stratosphere (12 to 35 km above the earth's surface) and damage and even destroy the earth's ozone layer. The ozone layer filters sunlight and absorbs parts of the UV radiation harmful to humans and other living beings. If the ozone layer is thinned by the effect of CFCs or even holes are created in it, this UV radiation increasingly reaches the earth's surface and leads to serious health consequences, which can range from skin damage to eye diseases, from weakening of the immune system to skin cancer. The discovery of the extent of the destruction of the ozone layer came as a surprise even to scientists and was a shock to the public and politics. This shock became even greater when it was predicted that it would take until 2050 to reverse the damage already caused to the ozone layer and restore its condition from 1970. The *ignorance* about the harmfulness of a specific human action led to a consistent international cooperation being quickly organized after its exposure, in order to reduce and then eliminate an existential threat to people on earth over a long period of time. Thanks to the assumption of political *responsibility* (see Chap. 6) by governments from all over the world, it was possible in 1989 in the form of a legally binding international agreement, the so-called Montreal Protocol, to first reduce and then stop the use and emission of CFCs.

All three examples illustrate how it is impossible to foresee certain dynamics and their consequences even in the relatively limited field of environmental and climate policy, and thus to *know* which challenges need to be overcome and how the political landscape will develop.

Understanding the Four Forms of Ignorance

- **Individual Ignorance:** This refers to a single person's lack of knowledge about certain facts, subject areas, and contexts. Individual ignorance can be partially mitigated through individual learning and further education.
- **Collective Ignorance:** This refers to a group, community, or society's lack of knowledge about certain facts, subject areas, and contexts. Collective ignorance can be partially mitigated through science and research.

- **Reducible Ignorance:** This refers to the ignorance of an individual or a collective that can be overcome through learning, further education, science, and research.
- **Irreducible Ignorance:** This refers to the ignorance of an individual or a collective that cannot be reduced even through learning, further education, science, and research. Reasons for this can lie in the complexity of a particular issue (e.g., the exact prediction of the weather in two years) or be of an epistemological nature.

This general description illustrates the difficulty of acquiring the necessary knowledge on certain issues when dealing with complex relationships. Such complex relationships often occur in the following contexts:

- In the interaction of different people,
- in the interaction between humans and nature,
- and in the interaction of different people or the interaction between humans and nature over long periods of time.

In these cases, it is often impossible to create a complete knowledge base. But why is this problematic?

When making decisions, whether as a private individual, in our profession, in voluntary work, or in politics, we usually try to select the best possible course of action using as comprehensive information as possible. We refer to these as *informed decisions*. This type of decision includes all important facts and carefully weighs different courses of action. Informed decisions are particularly important when dealing with pressing problems, where a good solution has positive effects for many people, while a poor solution leads to negative consequences for those affected. Our decision then not only affects us, but we also bear *responsibility* for others and must proceed with utmost conscientiousness.

To illustrate, consider persons in public administration responsible for public waste management and overseeing a large landfill on the city's outskirts. This landfill has been used for many years to dump the city's waste, separate recyclable materials, and deposit non-recyclable waste as safely as possible. The persons in our example have only been responsible for the landfill for a few years, but carry out their work conscientiously and base their decisions on broad technical expertise. Now, imagine that during an inspection, a leak is discovered where wastewater from the landfill threatens to escape and enter surrounding streams, posing a health risk to people and other living beings in the area. In this situation, it is crucial for our responsible persons to gather as much knowledge as possible about the situation, the potential dangers, and different solutions. Even though we will argue below that we need to consider our *ignorance* more in many decision situations because we usually cannot obtain a complete knowledge base, this does mean that every effort should be made to gather as much and as reliable knowledge as possible. Only in this way can we arrive at a well-informed decision. In our example, it would be essential to gather the available and relevant knowledge and take it into account in the solution proposals. So, for example, the following questions would need to be answered:

- What types of waste are stored in the landfill?
- What interactions can occur between the substances in the landfill?
- What leaking toxins can be expected?
- What is the technical design of the landfill?
- At which points could a leak have occurred?

These and other considerations then lead to the development of a solution proposal that best addresses the problem.

Here we arrive at our main point, which directly follows the emphasis on the necessity of gathering as much reliable information, as much knowledge as possible. The message in connection with knowledge and ignorance is that even our conscientious protagonist at the landfill must consider that even a meticulous compilation of relevant information cannot prevent her from remaining ignorant about certain aspects, relationships, and developments. However, these unknown aspects, relationships, and developments could be of central importance for our decision regarding a solution.

In our example, it is conceivable that several years ago, highly toxic chemical waste was illegally dumped in the landfill unbeknownst to those in charge. This fact could not have been taken into account early on to adjust their approaches accordingly. There is complete ignorance about the possibility that much more could be released through the leak than the toxins that were calculated based on the known waste composition. The persons in charge must respond to the leak as quickly as possible. The fact that they could not have known about the illegal chemical waste does not change the urgent need to find an adequate solution to the problem, even though they cannot fully understand the problem.

To what extent does the message we emphasize, which can be summarized with the slogan "consider your ignorance", help at this point? Specifically, this message suggests that we should abandon the idea of designing "perfect solutions" and "master plans" based on our knowledge. This assumption often leads to overconfidence in our cause and our plans become rigid. Once decided, adjustments are often no longer planned. Instead, we should design plans that are flexible and enable us to act even when *unexpected developments* occur that we could not foresee, which implies that, regarding their possibility, we were completely ignorant.

For our example, this means that it is very helpful for the person in charge to prepare for possible plan changes when working out a solution. They have consulted experts for the toxins calculated based on their knowledge and found out where they can get appropriate chemical binders to neutralize the known toxins. But because they did not solely rely on their knowledge and considered that their knowledge base might appear complete, but could in reality be incomplete, they have taken appropriate measures. And so, thanks to an extensive pollutant measurement at the landfill leak that they additionally ordered, it is noticed that toxins are leaking that should not actually be present based on the calculations. They wouldn't have needed this pollutant measurement at all—they seemed to know exactly which toxins could result from the waste composition known to them. The fact that they maintained an open attitude when developing solution proposals, despite their knowledge, enables them to deal with the unexpected information, namely that completely different and much more dangerous toxins are leaking from the landfill, despite other calculations and expectations.

They can flexibly deal with this *novelty*, adjust their plans and find adequate solutions. If they had been sure of their cause and had trusted their supposed knowledge base, they might now be standing at the leak, unable to act with inadequate chemical binders. Or, even worse, they would have fixed the leak in a way that would have been adequate for the calculated toxins, but unable to bind the much more harmful toxins. These could then have continued to leak unhindered and without the knowledge of those involved from the landfill and could have polluted the environment. The fact that this did not happen to our protagonists is due to their awareness of their own ignorance, their openness to unexpected developments, and their *power of judgement* (see Chap. 7) in weighing up possible courses of action.

12.1.2 Becoming Aware of Ignorance—Remaining Capable of Action

Building on this example, we aim to provide a definition that makes the complex term of ignorance more operational:

Important to Know: Definition of Ignorance
We define ignorance as a state where neither the possible outcomes nor their probabilities are known.

What does it mean, in practical terms, to become aware of our ignorance and to consider it more seriously? In the example given above, we analysed a case in which ignorance presented itself as (at least temporarily) *non-reducible ignorance*. The person responsible for the dump could not have obtained information about the illegally dumped waste in their situation. They were literally and completely blamelessly *ignorant*. The example also clarified that ignorance initially seems to deprive us of the basis on which we depend for our decision-making: We cannot base the decision entirely on knowledge, nor on calculable probabilities or an overview of possible outcomes. There is no other choice but to make a decision based on incomplete information with the help of openness and *power of judgement*.

In fact, there are many other situations in which we are not completely powerless in the face of ignorance. Ignorance is often *reducible*. The question is only how this can be achieved. The basis for this is that we, as individuals, as a group, or society, are aware of our ignorance, we then refer to this as *open ignorance*. As individuals, we strive to reduce this *individual ignorance* by learning and continuing education. As a group or society, we can reduce our *collective ignorance* by promoting science and research. It is important to note that despite all efforts, there will always be areas that remain unknown to us. And this is not only because we are dealing with hidden operations and deliberately concealed facts, as in the example of the dump. Anyone interested in the history of science will observe an interesting simultaneity of knowledge and ignorance: Every groundbreaking discovery opens the door to new questions and points us to new areas in which we are ignorant. Figuratively speaking, we pass through the door to knowledge and encounter new locked doors in the room beyond, which need to be opened through new research.

The good news: Those who are aware of their ignorance and take it into account remain capable of action—either because they can begin to gain new knowledge through education and research, or because they design plans with the appropriate openness, an openness that anticipates surprises and new developments and allows for flexible reactions when they occur. It is particularly problematic when we are dealing with ignorance of which we are not even aware and of which we are not conscious, so-called *closed ignorance*. This type of ignorance catches us completely unprepared and, so to speak, on the "wrong foot".

Important to Know: *Open* and *Closed Ignorance*

We distinguish between two further types of ignorance:

- **Open Ignorance:** We are aware that we are ignorant in a certain area and take this fact into account. In the case of reducible ignorance, we strive to reduce the ignorance by learning and researching. In the case of non-reducible ignorance, we include this fact in our action considerations and planning.
- **Closed Ignorance:** We are unaware that we lack knowledge in a certain area and are ignorant about it. For this reason, we neither plan measures to reduce the ignorance, nor do we take the possible ignorance into account in our action considerations and planning.

Our concern in this chapter is to avoid exactly this in our society, which supposedly has more and more knowledge and creates ever better access to knowledge. However, even the invention of Wikipedia and Explainity videos will not change the fact that we will never completely overcome our ignorance, neither individually nor collectively.

It should have become clear why 40 years of research on environmental problems has resulted in ignorance becoming more and more important to us. To quote the words of the economist and Nobel laureate Friedrich August Hayek: "It is high time that we take our ignorance seriously!" (1972, p. 32).

12.2 Defining Ignorance: When We Can and Cannot Speak of It

We have examined the phenomenon of ignorance using various examples. It may already be clear that in many instances where we believe we are ignorant, a closer look reveals that we do indeed know quite a bit about the situation. We stress that a more robust consideration of ignorance is necessary. However, it is also crucial to clearly distinguish when we can accurately refer to ignorance and when we cannot. We propose a more precise classification of unexpected events, developments, and facts, distinguishing between *risk, uncertainty*, and *ignorance*.

We will explain these three forms using an example (see Faber & Proops, 1998):

Key Distinctions: *Risk, Uncertainty* and *Ignorance*

- **Risk:** We refer to risk when we know both the possible events and their respective probabilities. A prime example is a dice game with a standard six-sided dice. We know all possible outcomes—the numbers 1,2,3,4,5,6 can appear. And we know the probabilities of these events, each of which is 1/6.
- **Uncertainty:** We refer to uncertainty when we know the possible outcomes, but do not know the probabilities of their occurrence. An example is earthquakes. We know, for instance, that earthquakes of varying magnitudes can occur in San Francisco due to tectonic conditions. What we do not know are the probabilities and the magnitudes of these potential earthquakes.
- **Ignorance:** We refer to ignorance when neither the possible outcomes nor their probabilities are known. An example is the damage to the ozone layer caused by the emission of chlorofluorocarbons (CFCs) on Earth. Until the theoretical possibility was discovered in 1974 and the actual ozone hole was observed in 1985, no one knew that CFCs could have this effect on the ozone layer or the probability of this occurrence.

Consider a person who frequently attends horse races and enjoys betting on the victory of various horses. There are two factors to consider: i) Any participating horse can win and ii) the (subjective) probability of winning varies from horse to horse. This probability is based on the spectator's subjective experience—"this horse has won many races"—and perhaps on his intuition about the form of a certain horse on a particular day. In the case of a bet, our spectator knows all possible outcomes and can assign a certain probability to each. When all possible outcomes are known and a subjective probability can be assigned to each, we refer to this as *risk*.

Now, imagine a second scenario. Since our racing fan does not regularly check the weather reports, he may arrive at the racetrack only to find, to his surprise, that the race has been cancelled due to bad weather. The outcome "No race takes place" is theoretically known to him as a possibility, but he has not assigned a subjective probability to this outcome, so he has, proverbially, "not counted on it". When possible outcomes are known, but no subjective probability is assigned or can be assigned, we refer to this as *uncertainty*.

Now, let us imagine a third scenario. After a long absence from his hometown, our racing enthusiast returns and resumes his hobby. He goes to the racetrack, or rather where he expects the racetrack to be. But instead of the racetrack, he finds a supermarket there, much to his surprise. He not only did not expect this outcome, he did not even consider it possible that someone would replace the racetrack with a supermarket. In this case, we refer to the outcome as being simply *unknown*.

The example illustrates that the lack of knowledge we encounter in different situations can indeed present itself in various ways. Sometimes we do not know anything specific, but we can gain an overview of the situation by thinking carefully. For instance, by considering the outcomes that might occur *(risk* and *uncertainty)* or when we even determine the possible outcomes and their probabilities *(risk)*. This information can help us weigh up different courses of action. Only ignorance leaves us in the dark in this respect—we then have to make decisions even though we have neither secure knowledge, nor an overview of the conceivable possibilities, and no probabilities available. In situations where decisions must be made against the backdrop of ignorance, *power of judgement* is required (see Chap. 7).

12.3 Overestimating Knowledge and Underestimating Ignorance

So far, it has become clear that we need to be aware of how often we are confronted with ignorance and how frequently we overestimate our knowledge. This understanding is crucial because it enables us to find a way to manage our ignorance and not be continually surprised by unexpected developments. Here is an example:

The mountain village of Brienz is situated in the Swiss canton of Graubünden, near the city of Chur, in the heart of the Swiss Alps. Above the village, mountains tower, and directly behind it lies a section of the mountain that has gained fame as the "Brienz Slide". In recent years, various media outlets have referred to it as the "best monitored mountain in Switzerland". The rock masses there have been meticulously monitored using the latest technology. The reason for this is that a part of the slope, known as "the island", has been moving for years. It was clear that these rock masses would eventually slide down into the valley in a landslide.

On 16.05.2023, the Tagesanzeiger headlined: "The mountain is coming—but how?"

Brienz has been struck by landslides repeatedly in the past. Between 1878 and 1907, a powerful debris flow moved approximately one meter daily. In 2019, a 100-ton rock hurtled past a playground at nearly 100 km per hour and came to rest in a meadow near the village. To protect the village's population and to provide timely warnings, the slope, which regularly moved about 1.9 million cubic meters of rock, was monitored around the clock by a team of experts. Five technical systems, consisting of GPS/GNSS, photogrammetry, tachymetry, rockfall radar, and georadar, analyzed the slope day and night, monitoring over 90 measurement points. In the spring and early summer of 2023, the slope moved into the valley at a speed of 40 m per day.

At this point, the case becomes interesting for our work with the category of ignorance. Scientifically and technically, everything possible was done in Brienz to understand as many variables as possible and to gain comprehensive knowledge about the processes on the mountain. The responsible parties could determine at any time which area of the slope was moving and at what speed. However, until the actual event on the night of 15.06. to 16.06.2023, no one knew exactly how the rock masses would loosen and what consequences the landslide would have for the village of Brienz. Even with the excellent available data, only scenarios could be formulated and probabilities named (Laukenmann, 2023):

1. The first scenario, a rockfall with a break-off of 250,000-500,000 cubic meters, was estimated with a probability of **60 %**. In this scenario, the northern part of the village might have been hit by falling masses.
2. The second scenario, a debris flow, in which the entire "island" with its 1.9 million cubic meters would slowly slide into the valley, was estimated with a probability of **approximately 30 %**. The concrete effects of this scenario remained a question of probabilities. It could not be clearly stated whether the area in front of the village, the entire village, or an even larger area would be hit.
3. The third scenario involved the danger of a landslide, in which the entire island would thunder into the valley. The probability of this scenario was given as **10 %**. Even for this case, the concrete consequences and the affected areas could only be approximately named using probabilities.

In fact, "the island" finally slid off between 23.00–24.00 hours on 15.06.2023 as a fast debris flow, which transported 1.2 million cubic meters of rock into the valley and stopped just a few meters in front of the schoolhouse on the outskirts of the village of Brienz (Joller, 2023).

With regard to ignorance, the example of Brienz is particularly impressive because even a team of experts working around the clock cannot completely overcome the ignorance about the actual interrelationships of the relevant systems and make reliable predictions. This is true even when, as in our example, almost perfect conditions are present: In Brienz, the analysis was of a spatially limited area, for which excellent data were collected from five complementary "state of the art" measuring systems. These data were then fed into established and tested models. Equally impressive is how the responsible parties in Brienz handled the situation: Despite the many available facts, they were aware of their ignorance and did not let the technology deceive them about the limits of their knowledge. Only because of this attitude was the population evacuated in time, no promises were made regarding the future of the village and, perhaps most importantly, the temptation was resisted to bring the slope down in a supposedly "controlled" manner by blasting it.

Summary: The Benefits of Acknowledging Our Ignorance

- **Firstly**, acknowledging our ignorance fosters an attitude of openness and flexibility towards unexpected events, as described in the first part of the chapter. This awareness allows us to act even when we lack comprehensive information for our decisions.
- **Secondly**, being aware of our ignorance heightens our attention to developments that may initially seem incidental. When we view *ignorance* as a structural component in the consideration of environmental problems, we can implement specific preventive measures. We can question the relationships and consequences of our actions (see Chap. 13 *Joint Production*) that we may not fully understand, and pay close attention to whether unexpected developments and *novelties* arise over time following our decisions. Furthermore, we can intentionally seek knowledge in areas that are not our primary concern, to prevent unforeseen problems from emerging later. In this second point, acknowledging our ignorance not only protects us from losing our *openness* to unexpected developments but also enables us to take *preventive steps* to reduce our *ignorance*.

References

Faber, M., & Proops, J. L. (1998). *Evolution, time, production and the environment*. Springer Science & Business Media.

Hayek, F. A. (1972). *Die Theorie komplexer Phänomene*. Walter Eucken Institut.

Joller, S. (2023, Juni 16). *Bergsturz Brienz—Warum Brienz verschont blieb*. Schweizer Radio und Fernsehen (SRF). https://www.srf.ch/wissen/bergsturz-brienz-warum-brienz-verschont-blieb.

Laukenmann, J. (2023). *Drei Szenarien für Brienz: Der Berg kommt – nur wie?* Tages-Anzeiger. https://www.tagesanzeiger.ch/der-berg-kommt-nur-wie-886815758799.

Joint production: When We Produce Not Only What We Want to Produce

13

Contents

What's it About?

This chapter discusses joint production. Every production of a material good leads to a simultaneous production of so-called *joint products,* which are often harmful to the environment. With the rapid economic growth since the industrial revolution, the amount of harmful joint products has also increased, causing severe damage to humans and the environment to this day. This was possible because the creation of joint products was long considered incidental.

M. Faber et al., *Sustainable Action in Economy and Society*,
https://doi.org/10.1007/978-3-662-69122-9_13

Traditional economics considered joint production, but only after it had already occurred, for example, when a river was already polluted. The approach presented in this chapter takes a different path: We demonstrate that joint production a) is thermodynamically conditioned and b) therefore must necessarily be considered before any production takes place, so that environmental problems do not arise in the first place.

13.1 Introduction to the Concept

Environmental policy approaches often do not lead to satisfactory results, as a fundamental characteristic of human-environment relationships is not sufficiently taken into account. Every action, especially every production and consumption process, requires energy. For thermodynamic reasons (see Chap. 9 *Thermodynamics*), it follows that every action has more than one effect. For example, the combustion of coal for electricity generation not only produces electricity, but also a number of other effects, such as the emission of carbon dioxide (CO_2), wastewater, waste heat, dust, etc. This circumstance is conceptually captured by ecological economics as the phenomenon of *joint production*. The (side) effects that arise in addition to the intended effects of an action are referred to as *joint products*.

In contrast to this fundamental view of ecological economics, there are prominent approaches to investigating environmental problems that use a disciplinary perspective.

- In philosophy, Hans Jonas coined the term "heuristics of fear" (see Chap. 6 *Responsibility*),
- in legal studies, there is consideration of endowing nature with fixed "rights", and
- in environmental economics, attempts are made to "internalize" negative impacts on the environment, i.e., to take them into account, for example, in the pricing of goods (taxes, levies, subsidies).

Despite all the merits of these approaches, they are not capable of simultaneously considering natural science and social science perspectives and systematically linking them. The concept of joint production allows this. It is thermodynamically based, applicable to production and consumption, and points out harmful effects of the economy on the environment and thus also on society. Taking into account joint production in environmental policy approaches allows us to recognize possible impacts on the environment before an action is carried out and to include them accordingly in the considerations. For these reasons, the concept of joint production is a central component of ecological economics.

In the following, we will systematically develop the approach of joint production. With this approach, we create a foundation, a so-called "Archimedean point"[1], from which we can start in the consideration of any environmental problems.

Our starting point is: Human existence is inconceivable without its relationship to nature. This relationship is a double one (Baumgärtner et al., 2006, Chap. 1).

1. On the one hand, humans are dependent in many ways on the services provided by nature. Examples include resources such as water, food, and fuels, as well as negative impacts such as floods, epidemics, and diseases. Added to this are cultural services such as recreation and the aesthetics of landscapes. Furthermore, nature provides a space for the unfolding of all kinds of human activities.
2. On the other hand, human action affects the natural environment. Humans change their natural environment, intentionally and unintentionally, to shape their living space. They extract resources from natural ecosystems; they return waste and pollutants to natural ecosystems and thereby change natural processes and functions.

The mandate of sustainability requires maintaining the functionality of nature and services of nature for humans in the long term (see Chap. 4 *Sustainability and Justice*). Therefore, the relationship between humans and nature must have a certain quality. How can this be ensured? This requires humans to adapt their behavior accordingly. But how? How can we approach this difficult and comprehensive question?

Firstly, we must acknowledge that there is a complex interplay of three distinct spheres at first glance. These spheres encompass the following processes:

- Processes within the sphere of nature, typically studied by the natural sciences;
- processes within the social, economic, and political sphere, traditionally analysed by the social and economic sciences;
- alongside these processes, there are categories of human thought, which are attributed to philosophy.

It is clear that the interplay of these three areas necessitates an interdisciplinary approach, which is a challenging and comprehensive requirement.

Our task is further complicated by the fact that our environment is constantly changing due to natural processes, developments, and human interventions. Similarly, our social, economic, political, and cultural conditions, as well as our environment, are continually evolving due to human action. This occurs over time (see Chap. 8 *Three Dimensions of Time* and Chap. 11 *Evolution*). Natural, societal, and economic processes are dynamic and characterized by changes in stocks

[1] "Give me a point where I can stand securely, and I will lift the world out of its hinges" (Archimedes, 285–212 BC).

(see Chap. 15), whether these changes are short-term, medium-term, or long-term. Furthermore, it is important to note that many of these processes are irreversible (see Chap. 10).

Since natural, economic, and social processes are inseparably intertwined, we are dealing with complex systems. This complexity is further exacerbated when we consider it in the long term. What are the reasons for this?

Knowledge about the relevant processes and systems is limited due to a high degree of uncertainty and fundamental ignorance (see Chap. 12 *Ignorance*). Human action is diverse and cannot be analytically captured with a single paradigm. The economic sciences' explanatory model for human behaviour is Homo Economicus (see Chap. 5 *Concepts of Man*), characterized by egocentric and optimizing behaviour. Despite growing criticism of this one-sided view in recent decades, it still plays a significant role in the social sciences and especially in the economic sciences. It is necessary to supplement this image of man with other images that, for example, take greater account of creativity, free will, and community spirit. It is clear how challenging it is to trace all these phenomena to their effects.

Faced with such a significant challenge, some might resign and claim that sustainability is a myth with little practical significance, while others might resort to isolated ad-hoc measures, which are insufficient to adequately address the challenge of sustainability.

In this book, we aim to take a constructive approach. What is actually necessary is a *comprehensive*, *consistent*, and *systematic* view of the problem and corresponding guidelines for action (Faber & Manstetten, 2003 [2010], p. 26 [pp. 8-12]). We have learned over more than four decades that it is indeed possible to develop guidelines for a sustainable policy and will demonstrate in the following that the concept of *joint production* is a central concept for the development of such a perspective (Baumgärtner et al., 2006).

What is the concept of joint production? Joint production occurs when the production of one good technically necessitates the simultaneous production of one or more other goods. These other goods can be desired or undesired. However, they are usually undesired. The reason for the climate problem is that the burning of coal, oil, or gas not only produces energy but also carbon dioxide (CO_2).

Important to Know: Human Action has Unintended Side Effects

The concept of joint production captures the unique feature of human action that it has unintended side effects; these are the structural causes of many environmental problems. Thus, the phenomenon of joint production is a natural starting point for investigating environmental problems and solving them in a sustainable way. It allows for a systematic and unified consideration of the problems mentioned above.

In the following, we aim to develop a comprehensive analysis of the interactions between economy and environment within the discipline of Ecological Economics, founded at the end of the last century. This designation already indicates that it is an interdisciplinary field. In terms of content, Ecological Economics is about the conceptual foundation of the science and management of sustainability (Costanza, 1992).

13.2 Joint Production and Thermodynamics

The insights presented in the chapter on *Thermodynamics* (Chap. 9) enable us to develop a comprehensive, scientifically based perspective on the relationship between the production and consumption of goods and the resulting environmental problems. Joint production is closely related to the laws of thermodynamics, especially the second law on entropy. It is important to reiterate that the two laws of thermodynamics have significantly contributed to the foundations of the discipline of Ecological Economics. How did this occur? The usefulness of thermodynamics arises because it underpins all economic activity; its two laws apply to all real production processes.

Important to Know: Thermodynamics as a Basis for Interdisciplinary Collaboration
Since thermodynamics is a fundamental basis of other natural sciences such as chemistry, biology, ecology, and engineering, a connection of Ecological Economics to other natural sciences is established. This facilitates interdisciplinary collaboration, a prerequisite for successful environmental policy.

In the chapter *Thermodynamics* (Chap. 9), we introduced the concept of an 'open system', which absorbs energy and matter and also releases both to its environment. From a thermodynamic perspective, the human economy is an open system embedded in the larger, but closed system[2] of the natural environment[3] (Ayres, 1978; Boulding, 1966; Daly, 1977; Georgescu-Roegen, 1971, and many more). The strength of the concept of joint production lies in its ability to examine the interactions between economy and environment within the framework of Ecological Economics. This can be demonstrated as follows: From a thermodynamic perspective, energy and matter are fundamental factors of production. Every

[2] A closed system only exchanges energy with its environment, but not matter.

[3] We consider our planet as a closed system, as energy is absorbed through solar radiation and released in the form of heat. The material exchange with the universe is negligible, as it is limited to occasional input of matter through meteorites and objects launched into space by humans.

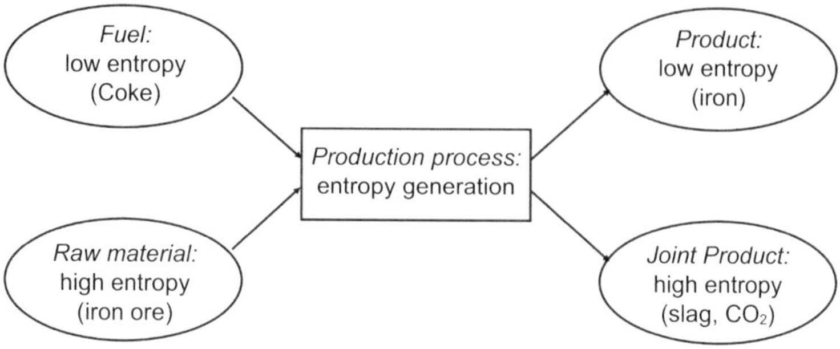

Fig. 13.1 Thermodynamic representation of iron production

production process is essentially a transformation of these factors. Consequently, the production processes are subject to the two laws of thermodynamics. This means that in an isolated system[4], matter and energy are conserved, as they can neither be created nor destroyed (1st law), and every real transformation process of energy generates a positive amount of entropy (2nd law).

In general, a production process can be described as a transformation of a certain number of inputs of production factors into a certain number of outputs of goods and pollutants, with all inputs and outputs characterized by their mass and their entropy. From the laws of thermodynamics, it follows that every production process is joint production, i.e., it necessarily leads to more than one output (Baumgärtner et al., 2006, Chap. 3; Faber et al., 1998). In particular, for industrial production processes that produce desired goods with low entropy, it is necessary and inevitable that joint products with high entropy are created.

We illustrate these consequences of thermodynamics for real production processes in Fig. 13.1. For example, in the production of iron, iron ore is used, which due to its mixture with earth—it is therefore contaminated iron—has a high entropy. To produce iron, which has a low entropy, the entropy of the starting material iron ore must be reduced. This can be achieved by supplying energy with a raw material with low entropy, e.g., with the energy carriers coal or coke. From a thermodynamic perspective, a shift of the high entropy of the iron ore into the waste products CO_2, slag, dust, wastewater, etc. is thus carried out to produce high-quality iron—with low entropy.

Figure 13.1 shows a typical industrial production process using the example of iron production.

[4]An isolated system does not exchange either energy or matter with its environment.

13.3 Every Production is Joint Production

The concept of joint production captures the essential thermodynamic restrictions regarding production processes, as dictated by the first two main laws. The diversity and generality of the joint production concept are evident in the fact that it encapsulates four central themes of ecological economics (see Baumgärtner et al. 2006: 5):

- Irreversibility (see Chap. 10),
- Limits of substitution of production factors,
- The ubiquity of waste, and
- The limits of growth.

These themes will be explained as follows:

- Irreversibility is explicitly included in the thermodynamic formalization of joint production mentioned above, as every production process generates entropy and is therefore irreversible.
- Limits of substitution are also considered, as the requirement must be met that materials with high entropy, such as iron ore, must be converted into desired goods with lower entropy, like iron. This presupposes that an irreducible minimum of low entropy fuels must be present.
- The ubiquity of waste follows from the fact that the production process necessarily generates entropy, which is contained in unwanted material and therefore represents waste (e.g., CO_2, slag, etc.).
- The combination of the three aspects mentioned above leads to the concept of the limits of growth, underlining the comprehensive explanatory value and the universality of the concept of joint production for understanding the economy in general.

Important to Know: Joint Production as a Concept for Interdisciplinary Work on Environmental Problems

The term joint production provides a simple and easy-to-use concept for investigating environmental problems in an interdisciplinary and comprehensive manner.

13.4 Joint Production vs. Theory of External Effects

In contrast to the approach used by ecological economics to investigate environmental problems with the concept of joint production, environmental economics of conventional economics uses a completely different approach: the *theory of external effects*. How do these two approaches differ? To understand this, it is necessary to first explain the term external effect. Consider a chemical company that discharges its wastewater into a large lake. Over time, this changes the water quality to such an extent that fishermen can catch fewer fish; their income decreases accordingly. However, the company does not primarily include the loss of the fishermen in its economic calculation, as it is not directly affected by it. The damage only occurs for "externals"—i.e., people not included in the economic model and is therefore not compensated for through the market. Therefore, the decline in fish is referred to as a (negative) external effect. There is thus an asymmetric relationship between the entrepreneur and the fishermen.

It usually takes a long time for the effect of such unequal relationships to lead to environmental policy consequences when the theory of external effects is applied. This will be explained using the example of the development of the water quality of Lake Constance after the 2nd World War. From 1950, it was increasingly polluted by wastewater from municipalities and the economy. The problem was addressed in 1959; but in particular, the phosphate input from fertilizer leaching continued to increase. It was not until the beginning of the seventies that sewage treatment plants were built. However, until 1979, the phosphate input still increased. It took until around 2010 for Lake Constance to regain its original water quality. The cost of building the water purification was several billion euros.

How do the explanatory approaches of ecological economics with their joint production approach and environmental economics with their approach of external effects differ? The former start from the *cause*, namely the creation of a joint product, in the mentioned example the leaching of wastewater and fertilizers into Lake Constance. In contrast, the latter starts from the *effect*, i.e., the degree of pollution of Lake Constance, at which politics is willing to take measures. As we have seen above, this leads to a time delay. We can observe something similar with the climate problem.

Important to Know: Difference between Ecological Economics and Environmental Economics

In summary, we find: Environmental economists only consider welfare effects when they have occurred. This means that external effects are a matter of hindsight, we can also say they are only considered ex-post. On the other hand, ecological economists, using the concept of joint production, draw attention to possible environmental damage in advance, i.e. ex ante. This is a strong incentive to consider unknown potential welfare losses already when introducing a new product, such as that of fertilizers.

13.5 The Case Study of the Soda-Chlorine Industry

The English proverb, "the proof of the pudding is in the eating", suggests that the value of an insight cannot be demonstrated through theoretical considerations, but only through practical experience. This proverb is also applicable to the field of human-environment relationships. We illustrate the relevance of joint production for the economy, society, and politics using a case study of the soda-chlorine industry, which we examine over a period of more than 100 years. In the mid-18th century, the British textile industry grew so significantly that the potash needed for bleaching, obtained from wood, was no longer sufficient due to a wood shortage: it had to be replaced by soda. However, natural soda became scarce at the end of the 18th century. Due to high demand, the French Academy of Science offered a prize in 1775 for the production of synthetic soda, which was eventually awarded to Nicholas Leblanc in 1791. Nevertheless, his process was not implemented in large-scale plants until 30 years later in England.

However, soda was not the only product produced. The environmentally destructive joint product, hydrochloric acid, was also produced. Consequently, people living near the factories and the neighbouring agriculture were adversely affected. It was not until 1863 that politics intervened by mandating that hydrochloric acid should no longer be released into the air through chimneys, but directly into bodies of water. This, however, led to not only fish dying but also ships and locks rusting. This resulted in a second change in the law in 1874, which prohibited this type of disposal.

A new development arose with the invention of the Deacon[5] process, which made it possible to obtain pure chlorine from hydrochloric acid. Since pure chlorine could be used for many purposes and found a brisk sale, an unwanted joint product could be turned into a desired one. However, over time it became apparent that this merely shifted the environmental problem from the sphere of production to that of consumption; because the disposal of chlorine-containing products leads to serious consequences, as the example of the creation of the ozone hole in the atmosphere later showed (Faber & Manstetten, 2014, pp. 245–248).[6]

Important to Know: Theory of Joint Production versus Theory of External Effects

The theory of external effects is not easy to convey to laypeople because it requires a considerable amount of economic theory. In contrast, the phenomenon of joint production is an easily understandable principle for

[5] Henry *Deacon* (1822–1876) was a chemist and industrialist.

[6] Further case studies on joint production in the production of paper, cement, and sulfuric acid can be found in Baumgärtner et al., (2006, Chap. 15–18).

understanding environmental problems. Precautions can be developed so that these problems do not arise or their effects can be reduced.

Due to the close relationship between joint production and the two main laws of thermodynamics, the concept is suitable for interdisciplinary cooperation, as already mentioned above. After all, the ubiquity of joint production leads to questions of responsibility (see Chap. 6).

13.6 Why the Concept of Joint Production is Important for Society's Handling of Environmental Problems

We will now explain what joint production specifically has to do with the economy and environmental resource problems. We have already seen that in the production of a good, one or more other goods are inevitably produced due to thermodynamic laws. We refer to this phenomenon as joint production. The additional outputs, so-called joint products, can be partly desired and partly undesired.

Economists and business economists have dealt with the economic aspects of joint production in various ways since the middle of the 18th century (Baumgärtner et al., 2006, Chap. 6). The reasons for this are varied. Economists examine macroeconomic aspects, while business economists focus on entrepreneurial aspects. Due to increasing environmental and resource problems, it is primarily the undesired joint products that are in focus both macroeconomically and in specific companies. However, it should not be overlooked that there are also many desired joint products, for example in the chemical industry (Müller-Fürstenberger, 1995).

The concept of joint production, due to its easy comprehensibility and applicability, can serve as a didactic tool to illustrate the thermodynamic relationships of the interaction between the economy and the environment. This is because economists, business economists, entrepreneurs, and managers are much more familiar with the concept of joint production than with the effects of the laws of thermodynamics. The simplicity of this concept was explained in Fig. 13.1 using the example of iron production: In industrial production, environmentally harmful joint products are necessarily and unavoidably produced. This means that the production of a desired good is inevitably associated with environmental damage.[7]

[7] This is explained in detail in Baumgärtner et al. (2006) in Sect. 3.2 and physically proven in Sect. 3.3. In Sect. 3.4, it is shown what degree of thermodynamic (in)efficiency an industrial production process has and what quantities of waste are produced, and in Sect. 3.5, how these lead to environmental problems.

13.6.1 Thermodynamics of Joint Production

Referring to our representation in the concept of *Thermodynamics* (see Chap. 9) and the relationships explained in the introduction of this chapter, we will now expand and deepen our knowledge about joint production. We start with the following question: Since the joint products of a desired good are often unwanted and harmful to the natural environment, why is their occurrence not simply avoided? An answer to this question, which most economists would give, is that the occurrence of waste products and their disposal in many cases is a market inefficiency, which is due to an externality (see above), i.e. the consequences of waste production and disposal are not included in market prices. According to this economic view, the occurrence of waste is due to market failure, which in principle could be avoided by appropriate political measures, such as taxes or levies on waste.

While the problem is indeed to some extent due to a failure of the market (and politics), the view from a thermodynamic perspective reveals a completely different aspect. From a thermodynamic point of view, the occurrence of joint outputs appears as an unavoidable necessity of industrial production. Our example of iron production (Fig. 13.1) is of a specific kind, but the insights that can be gained from its analysis are applicable to all industrial production processes. Two essential thermodynamic results are:

1. Due to the main laws of thermodynamics, high entropy containing joint products in the industrial production of desired goods, where low entropy materials are used as energy sources, are unavoidable. This type of production is currently used worldwide and dominates the economic system in industrial economies.
2. The thermodynamic analysis allows quantifying the amount of waste products produced: All materials that enter the economy stay there, or leave in the form of waste. To get an idea of the amount of waste in the Federal Republic of Germany, we provide their amount in the year 2018: It was 417 million tons (Federal Statistical Office, 2020) or slightly more than 5 tons of waste per person.

There are three thermodynamic reasons why such large quantities and ultimately the waste of resources contained in the waste occur:

The first reason is the conservation of mass, which follows from the first law of thermodynamics. Starting from a raw material, which is a mixture of different chemical elements, to produce a desired product, which consists only of a certain chemical element, inevitably results in a material waste of all other chemical elements.

The second reason is the use of a material fuel, which is characteristic of many currently used industrial production technologies. The fuel—in the example of iron production carbon—serves only to provide the energy needed for the required chemical reaction, the carbon material itself is usually neither desired nor necessary in the reaction. Since the mass is conserved, the fuel must be discharged somewhere after its energy content has been subtracted. It becomes waste, including CO_2. An alternative way to provide energy for production processes, which produces significantly less harmful joint products, is the use of electricity from renewable energy sources, such as solar, wind, tidal, or hydropower.

The third reason pertains to the thermodynamically inefficient performance of current technologies in energy conversion. As highlighted in chapter 9 on *Thermodynamics* (see Section 9.6), the first steam engine invented by James Watt only utilized 1% of the energy, leaving 99% unused. This corresponds to a thermodynamic efficiency of 1%. It took a significant amount of time to increase the efficiency of steam engines, which currently ranges between 30 and 50%. Due to technical circumstances, heat energy is often generated but cannot be used, resulting in a theoretical maximum efficiency that is not only less than one but often considerably so. A 100% efficiency is impossible due to the 2nd law of thermodynamics, but substantial increases can be achieved. The inefficiencies of technology lead to higher energy consumption than is thermodynamically necessary, and consequently, the amount of material fuel and waste produced is greater than necessary. This is particularly true for carbon dioxide emissions when carbon (e.g., coal or coke) or hydrocarbons (e.g., petroleum or natural gas) are used as fuel. Despite the fundamental understanding that joint production is an inherent feature of industrial production technologies, there is significant potential for waste reduction. Many joint products can be repurposed in the economy, often through energy-intensive recycling. Moreover, a large portion of joint products is still produced due to inefficiently designed production methods.

Important to Know: Why Thermodynamics is Crucial for Understanding Environmental Problems

- Conceptually, the integration of joint production into thermodynamics is vital for ecological economics. This is because the thermodynamic concepts of matter, energy, and entropy form the scientific basis for all processes in **a)** the economy, **b)** the environment, and **c)** the interactions between these two.
- Thermodynamics provides a unifying perspective on the interaction between the biogeophysical environment and the economy. This scientifically grounded approach, combined with economic analysis, enables the posing of questions that would not be possible from the perspective of a single scientific discipline, such as conventional economics.

13.7 The Challenges of Thermodynamic Arguments

The field of thermodynamics is notoriously complex, particularly the concept of entropy, which even natural scientists often struggle to apply correctly. This complexity is frequently underestimated, leading to erroneous thermodynamically justified statements and claims. Such errors often result in unproductive discussions, especially within the environmental sector. A common mistake is the failure to adequately consider the type of thermodynamic system being used in an argument. For instance, conclusions drawn from open systems cannot be applied to isolated or closed ones. Any oversight or inaccuracy can lead to false conclusions. This is akin to navigating a moorland; it requires a knowledgeable guide who knows the safe paths hidden by the moor and can lead others safely through the terrain. A small misstep by someone unfamiliar with the moor can be life-threatening. Similarly, in the environmental field, such misjudgements can exacerbate problems, waste crucial resources, or create new issues.

It is unrealistic to expect everyone involved in science, universities, schools, authorities, politics, international organizations, and the economy to be well-versed in thermodynamics. It is already a significant expectation for basic knowledge to be present. Therefore, the question arises in ecological economics: how can we convey the central insights of thermodynamics, which are essential for solving environmental problems, without oversimplifying? In our research, teaching, public relations, and environmental policy advice, we have found it useful to communicate these insights through the thermodynamically grounded concept of joint production, which is more tangible and easier for others to comprehend.

The concept of joint production encapsulates essential thermodynamic insights about production and ecological processes:

- the conservation of mass and energy,
- the generation of entropy and thus,
- irreversibility.

In this context, the concept of joint production serves as a "translation" of fundamental thermodynamic laws into the language of ecological economics. It is useful for:

- formulating questions,
- and developing solution proposals.

Excursion: Karl Marx and Joint Production

Although different aspects of joint production have been studied for over 250 years by pioneers of economics such as Adam Smith (1723–1790), Karl Marx (1818–1883), William Stanley Jevons (1835–1882), Arthur Cecil Pigou (1877–1959), and John von Neumann (1903–1957), the significance of joint production for environmental policy was only recognized at the end of the 20th century. Until then, the still predominant, but as explained above deficient theory of external effects was without alternative. This led to an unsatisfactory ex-post consideration of environmental problems. A thermodynamically based theory of joint production has the potential to shift environmental policy to an ex-ante perspective and thus identify and address problems before they arise.

However, the approaches to joint production developed in the last centuries and still applied today lack a thermodynamic basis. We want to illustrate this using one of the great economic thinkers, Karl Marx, as an example. At various points in his work, Marx dealt with environmental and raw material problems in general, but also specifically with the phenomenon of joint production as a cause of waste, wastewater, and emissions (Petersen & Faber, 2018, pp. 141–145). He also discusses that it is expedient to recover raw materials through recycling due to price advantages. This view was unique for his time. However, due to his optimism about progress, Marx considered the problem of unwanted joint products to be only a temporary problem. He was confident that scientific progress, especially in the field of chemistry, would lead to the transformation of useless waste into desired products. He gives examples of how medicines can be obtained from tar. He trusted entrepreneurs to eventually operate every production process in such a way that no pollutants would be produced for the environment. Marx thus saw the phenomenon of joint production, but he considered it a problem that would be solved by the dynamics of the capitalist economy itself, as every waste could be recycled. But this optimism is not appropriate, as the laws of thermodynamics also set limits to the use of joint products. Many things are simply unusable due to their high entropy. This shows that Marx's understanding of joint production was not sufficient—a circumstance that can also be seen in later economists.

This incomplete understanding of joint production leads to false conclusions for the economy. For example, the thought that the growing amounts of waste just need to be recycled to solve the waste problem; because, that recycling has its limits follows from the thermodynamic foundation of joint production. If such a comprehensive understanding were to spread in society, it could contribute to the fact that it no longer orientates itself towards a naive growth idea.

13.8 Recognising and Understanding Environmental Problems Through Joint Production

We have examined the concept of joint production from two complementary perspectives: economic and thermodynamic. This interdisciplinary approach is based on theoretical considerations and examples. Many environmental problems stem from economic activity, which, for thermodynamic reasons, leads to unwanted joint production. Environmental pollution can thus be traced back to the demand for desired goods and their industrial production. In practical terms, understanding joint production allows us to recognise and address environmental problems at their inception, not only after the negative effects of production processes on humans and nature become noticeable. Consequently, it is suitable for realistically describing the relevant restrictions of economic action. This perspective forms the basis for a comprehensive sustainable environmental policy.

13.8.1 Joint Production in Ecological Systems

So far, we have primarily discussed economic systems, but we would like to emphasise that all that has been said also applies to ecological systems, as the phenomenon of joint production is ubiquitous there as well. As a rule, living beings, whether animals or plants, produce more than one output, be it a good or waste, simultaneously. Consider, for example, an apple tree; in addition to fruits, it provides leaves, which offer living space and shade for other creatures, and its bark provides a habitat for insects. Humans, like other animals, consume apples. An extremely important service, produced by all green plants as a joint product of their respiration, is the oxygen that other higher organisms need for breathing. Some natural communities, such as rainforests, even "produce" their own climate in interaction. Consequently, the climate in a tropical region is greatly altered when the prevailing rainforests are destroyed.

13.8.2 Joint Production and Interdisciplinarity

The concept of joint production is therefore not only applicable in the fields of economics and engineering sciences, but also in ecology and biology. This promotes interdisciplinary dialogue between scientists in these disciplines. Furthermore, we have found in teaching, in lectures in interdisciplinary committees, and in policy advice that environmental problems and corresponding solution proposals are easier to communicate with the concept of joint production than if this is attempted exclusively through the theory of external effects. This also applies in public lectures, whether at non-governmental organisations, churches, schools, or in media such as radio and television, because the concept of joint production and its implications are vivid and easy to convey.

13.9 Joint Production and Responsibility

A thorough understanding of this concept is a good prerequisite for developing successful environmental policy solutions, because it explains:

- why environmental problems arise, and
- how they can be reduced or even avoided.

We have already explained that the concept of joint production allows us to consider two very different perspectives simultaneously, namely that of natural science and economics. In Chap. 5 on *Responsibility*, we show that a third perspective emerges, namely that of ethics: Who takes responsibility for which joint products and their effects? The consideration of joint production should be supplemented by the question of responsibility for joint production in order to gain as comprehensive a perspective as possible on environmental problems. This provides a broad interdisciplinary approach to investigation, which is not limited to the view of a single discipline as is often the case with proposed solutions in environmental policy, and therefore does not neglect aspects outside the discipline.

References

Ayres, R. U. (1978). *Resources, environment, and economics: Applications of the materials/energy balance principle*. Wiley.

Baumgärtner, S., Faber, M., & Schiller, J. (2006). *Joint production and responsibility in ecological economics: On the foundations of environmental policy*. Elgar.

Boulding, K. (1966). *The economics of the coming spaceship earth.* In H. Jarrett (ed.), Environmental Quality in a Growing Economy, Resources for the Future/Johns Hopkins University Press.

Costanza, R. (1992). *Ecological economics: The science and management of sustainability*. Columbia University Press.

Daly, H. (1977). *The steady-state economy. The sustainable society: Implications for limited growth*. Praeger

Faber, M., & Manstetten, R. (2003). *Mensch-Natur-Wissen: Grundlagen der Umweltbildung*. Vandenhoeck & Ruprecht. [English Translation published 2010: *Philosophical Basics of Ecology and Economy*. Routledge.

Faber, M., & Manstetten, R. (2014). *Was ist Wirtschaft?: Von der politischen Ökonomie zur ökologischen Ökonomie*. Verlag Karl Alber.

Faber, M., Proops, J. L. R., & Baumgärtner, S. (1998). All production is joint production: A thermodynamic analysis. In S. Faucheux, J. Gowdy, & I. Nicolai (eds.), *Sustainability and firms, technological change and the regulatory environment* (131–158). Elgar.

Georgescu-Roegen, N. (1971). *The entropy law and the economic process*. Harvard University Press.

Müller-Fürstenberger, G. (1995). *Kuppelproduktion: Eine theoretische und empirische Analyse am Beispiel der chemischen Industrie*. Physica-Verlag.

Petersen, T., & Faber, M. (2018). *Karl Marx und die Philosophie der Wirtschaft: Unbehagen am Kapitalismus und die Macht der Politik*. Karl Alber.

Statistisches Bundesamt. (2020). *Abfallaufkommen in Deutschland 2018 bei 417,2 Millionen Tonnen*. Statistisches Bundesamt. https://www.destatis.de/DE/Presse/Pressemitteilungen/2020/06/PD20_195_321.html.

Absolute and Relative Scarcity: There is no Planet B

14

Contents

What's it About?

This is about scarcity. Scarcity explains why the overuse of raw materials, the overexploitation of agricultural land, polluted waters, the creation of waste, or the deforestation of rainforests become a problem for humanity. The simple answer is: **"There is no Planet B!"**

Thus, scarcity demonstrates that through its economic and lifestyle choices humanity is destroying its natural basis of life on a global scale. We, as humanity, do this even though we do not have unlimited possibilities to resort to alternative areas and resources, that is, even though we are confronted with scarcity. This chapter discusses the different forms of scarcity, how they change in the course of ongoing environmental destruction, and how we can deal with scarcity.

M. Faber et al., *Sustainable Action in Economy and Society*,
https://doi.org/10.1007/978-3-662-69122-9_14

14.1 Introduction to the Concept

The fact that we cannot switch to a replacement planet underscores a fundamental understanding of the concept of scarcity. In the context of economy and environment, we speak of scarcity when a resource, such as water, is not unlimited and freely available for our purposes. This means that our access to this resource is restricted. Scarcity can take two forms:

1. Relative Scarcity
On one hand, a resource can be relatively scarce, which always establishes a reference to other resources. This means that the acquisition or consumption of the resource requires the renunciation of other resources. For example, filling a pool with water usually costs money, so one must forego the consumption of other resources (e.g., a visit to the cinema). However, a person can freely decide whether to fill the pool or prefer another resource instead. The resource—water in the pool—is thus substitutable. The concept of *relative scarcity* therefore refers to substitutable resources, for the consumption of which other resources must be renounced. It presupposes freedom of choice in consumption.

In mainstream economics[1], this relative concept of scarcity is usually used when talking about scarcity.

2. Absolute Scarcity
In addition to relative scarcity, a resource can be absolutely scarce. This means that the availability of the resource is limited and the resource cannot be replaced by other resources. For example, the water brought along on a hike through the desert is absolutely scarce. For the hikers, who must drink to avoid dehydration, the trade-off between the water and any other resource no longer arises: water is no longer substitutable by other resources for them. Furthermore, the availability of water is absolutely limited on the journey through the desert. Consequently, the non-substitutability and the limitation of a resource mean that it is absolutely scarce.

14.2 The Development of the Concept of Scarcity in Economics

The British economist Thomas Robert Malthus (1766–1834) used the concept of scarcity to gain insights for a new understanding of the relationship between nature and man. His thesis was that the population always grows substantially

[1] By mainstream economics, we mean the branch of economics that predominates at universities today and in the 20th century and is predominantly used for decision-making in the economy, politics, and society. In contrast, ecological economics, plural economics, Marxist economics, feminist economics, and other economic perspectives have only a minor influence.

stronger than food production—creating a situation of scarcity—which in turn regularly leads to famines and population decline. Although Malthus' thesis was historically refuted, his approach contained a fundamental observation: natural resources are finite, setting a physical limit to economic growth based on increasing resource consumption.

This view of the scarcity of natural resources lost more and more importance in the further development of economics and was replaced by a more abstract concept: that of relative scarcity. This idea of scarcity moved away from human dependence on natural resources and was relativized by the focus on human needs and desires regarding substitutable goods.

Due to the ongoing consumption of resources and increasing environmental problems, the recognition of finite resources and the limited capacity of the environment to absorb pollutants experienced a resurgence beginning in the 1970s that continues to this day. Economists of neoclassical environmental economics as well as the newly formed ecological economics, real world economics, post-growth economics, and others recognized the limits to growth. The impossibility of escaping the scarcity of natural resources, perceived by Malthus two centuries earlier, was taken up again and captured by a founder of ecological economics, Herman Daly (1938–2023), under the term *absolute scarcity* (Daly 1977, p. 39).

The understanding of scarcity was thus initially absolute, then relative, and today it is both relative and absolute, depending on the respective discipline, school of thought, and perspective.

14.3 Economics and Relative Scarcity

We will examine in more detail when, from the perspective of mainstream economists, the scarcity of goods[2] arises and why scarcity in this view is relative.

Whether a good is scarce in the economic sense depends on

1. its availability, and
2. the subjective preferences of consumers regarding the good.

Regarding 1. For a good to be scarce, its availability must be limited. It is irrelevant whether the limitation is due to a restriction of occurrence, or costs incurred in the acquisition of the good—for example, money, physical effort, or time.

This can be illustrated using the example of the good "clean air". In rural areas, clean air is usually unlimitedly available and the appropriation of air, i.e., breathing, does not incur any costs. Therefore, clean air is not scarce for rural inhabitants. The situation is different, for example, for residents of Dhaka, the capital

[2] Under the term goods, we include both consumer goods and services provided by humans and ecosystems.

city of Bangladesh. Due to particulate matter pollution, the availability of clean air is locally severely restricted. To "consume" clean air, those affected have to drive to the countryside or clean the air in their apartments with suitable devices, both of which involve costs (time and money) for them. Thus, the good, clean air, can become scarce for them.

Regarding 2. In addition to availability, i.e., supply, the scarcity of a good depends on the subjective needs and desires of consumers, i.e., demand. A good can only be scarce for a person if this person is also interested in acquiring the good. Every Dhaka resident who values clean air is affected by scarcity regarding clean air. The situation is different for the conceivable case of a Dhaka resident who has no preference between fresh and particulate matter polluted air. Since she does not care which air she breathes, clean air is not a scarce good for her, regardless of its availability.

What makes the scarcity of goods just described relative? So far, we have only spoken of scarcity per se. A good becomes *relatively scarce* when the person concerned has the choice of whether to consume the good or prefer another good instead. *Relative scarcity* is therefore a question of choice between goods, or the substitutability of goods. In our example, the Dhaka resident may freely decide whether to drive to the countryside to breathe fresh air or not. Fresh air is therefore substitutable for her, she can choose another good instead. Instead of investing money and time in the trip to the countryside or in a device to clean the air in her apartment, she may enjoy the cultural offer in downtown Dhaka and acquire a gym membership to do something for her physical fitness. Therefore, her scarcity is relative to alternative goods.

Now we can ask: What if a good is not substitutable? Does the concept of *relative scarcity* used in mainstream economics foresee such cases? How should they be dealt with?

14.4 Economics and Absolute Scarcity

Mainstream economics assumes choices between goods, leading to the concept of *relative scarcity*. However, what if clean air, for instance, is not substitutable for someone? What if, in the case of severe lung disease, fresh air must be breathed (demand side), and the supply of clean air cannot be increased because even the air in the countryside is already polluted (supply side)? Cases of this kind are usually outside the scope of mainstream economics. In contrast, ecological economists deal with such cases, which they categorise under the concept of *absolute scarcity*.

The scarcity of a good is absolute when it is not substitutable, that is, when people lose the choice to decide between the good and another, and are thus necessarily dependent on the good. Non-substitutability has two aspects:

1. Non-substitutability on the demand side.
2. Non-substitutability on the supply side.
   For a good to be absolutely scarce, substitutability must be restricted on both, the demand and the supply side.

1. Let us first consider the demand side. In economic thinking, a distinction is made between existential and imaginary needs (Faber & Manstetten, 2003 [2010], pp. 183-186 [135–139]). Existential needs include everything that is absolutely necessary to ensure human survival, such as eating, drinking, a roof over one's head, warmth, and basic health care. Limiting global warming to 2 degrees could also be argued to be an existential need of a society, as higher warming endangers other existential needs. Imaginary needs include everything that extends beyond existential needs, that is, everything that concerns human life beyond the threshold of mere *survival.*

The scarcity of a good that endangers the fulfilment of an existential need is absolute if in addition, no substitutability is possible on the supply side—as in the above example of the water bottle in the desert. With imaginary needs, on the other hand, it is appropriate to speak of *relative scarcity*, as we are not necessarily dependent on them, i.e., we can substitute in the demand side. While this relationship is quite clear in the case of a car trip to the sea, it is not so clear with other imaginary needs, such as the desire for education. Although not vital, access to education can be seen as indispensable for human dignity and thus, in the case of limited educational opportunities, be absolutely scarce. It cannot be definitively clarified here which goods fulfilling imaginary needs are to be classified as absolutely or relatively scarce. This must be discussed in each specific case, also from philosophical perspectives.

2. Non-substitutability of goods can also affect the supply side. In certain cases, it is possible to make goods available in sufficient quantities or to replace them equivalently. This can be achieved by applying certain technologies. For example, the absolute scarcity of water in the desert could be partially alleviated by building a well. However, every technology is subject to physical limits (see Chap. 9 *Thermodynamics*), which limit the possibilities for substitution. Furthermore, ecology teaches us that certain services of an ecosystem, such as those of bees or the sight of a beautiful landscape, cannot be replaced equivalently; one reason for this is that we often do not fully understand their functions in nature (see Chap. 12 *Ignorance*). Therefore, the absolute scarcity of goods can only be partially prevented by substitution on the supply side, and in some cases, this is not possible.

14.5 The Significance of the Differentiation between Absolute and Relative Scarcity for Dealing with Environmental Problems

Whether environmental problems are viewed from the perspective of relative or absolute scarcity significantly determines how the respective problems are dealt with. We will demonstrate and discuss the significance of the concept of *absolute* and *relative scarcity* using a practical application, and how classifying scarcity as relative or absolute affects the handling of the environmental problem under consideration. For this, we will refer to a topic at the centre of the climate debate: The global remaining budget of CO_2 that can be emitted to limit global warming to below 2 degrees.

We ask: Which form of scarcity—relative or absolute—is suitable for describing the problem? In the following sections, we will use the case of the remaining CO_2 budget for reaching global climate targets to provide answers to the following questions: How does the dominant understanding of scarcity in a society shape the handling of environmental problems? How can one's own understanding of scarcity help to identify solution spaces for environmental problems? However, we would like to point out that the following explanations do not claim to be comprehensive in the consideration of the example or in answering the two questions, as this would exceed the scope of this book.

14.6 Remaining CO_2 Budget for the 2 °C Target: Absolute or Relative Scarcity?

At the World Climate Conference in Paris, the 195 participating countries (+ EU) agreed on December 12, 2015, to limit global warming to well below 2 °C, preferably below 1.5 °C. As of the first draft of this text in January 2022, approximately 1066 gigatons of greenhouse gas emissions (hereafter GHG, calculated in CO_2 equivalents[3], hereafter CO_2-eq) could still be released without jeopardizing the achievement of the 2 °C target with high probability.[4] By the time of the final revision of the text a year later, it was only 1024 gigatons. This budget, at the current unchanged emission rate (approximately 1337 tons per second), will be exhausted in just over 24 years, i.e., by 2047. Since it is almost certain that

[3] CO_2 is indeed the greenhouse gas contributing most to global warming, but it is not the only one. Other greenhouse gases include methane, nitrous oxide, chlorofluorocarbons, etc. To standardize the emissions of these other gases, some of which have a higher greenhouse gas effect per unit of weight, they are converted into CO_2 equivalents. In the following, we always refer to CO_2 equivalents when considering the remaining budget of GHG.

[4] The currently remaining CO_2 budget for the 2 °C and the 1.5 °C target can be checked daily on the MCC's website: https://www.mcc-berlin.net/forschung/co2-budget.html.

significant amounts of GHGs will continue to be emitted after that, the 2 °C target would thus be missed.

How can we apply the concept of *absolute* and *relative scarcity* introduced above to the topic of the remaining GHG budget? What ambiguities and difficulties arise in this context, and what can we learn from it?

The emission of GHGs, upon which a large part of our current economy is based, can be seen as the consumption of a resource, or more generally, a good. What do we mean by that? The environment has the ability to absorb pollutants from economic activity, whether from production or consumption, and to transform them to some extent so that they are no longer harmful. This amount is called the absorption capacity of an environmental area. This absorption capacity is a good provided by the environment. In our example of the GHGs budget, only 1024 gigatons CO_2-eq of this good—GHG that can be absorbed by the environment without leading to severe climate changes—are available at the beginning of 2023. Obviously, the good GHG is in demand by humans and is—if we consider adherence to the 2 °C target as a premise to be adhered to for global economic activity—only available in limited quantities; we can therefore say that the good GHG is scarce.[5] But is the scarcity of GHG to be considered as relative or absolute?

14.7 Greenhouse Gas Emissions as a Relatively Scarce Resource: Emissions Trading

In the current political discourse, the approach to adhering to the GHGs budget is primarily characterized by viewing GHGs as a relatively scarce resource. This perspective forms the foundation for the instrument of emissions trading. For instance, through the EU Emissions Trading System, a limited number of emission allowances, known as emission certificates, are issued and can be traded on the European market. In this system, a company subject to emissions trading must redeem a corresponding emission certificate for every ton of CO_2 -eq it emits. It is important to note that certain industries receive a specific quota of certificates from the EU free of charge for competitive reasons. Depending on its demand for GHGs, a company can

- emit GHGs and thus use up its potentially existing quota of certificates,
- purchase additional certificates to be permitted to emit more GHGs, or
- sell unneeded certificates.

---

[5] Even if we were to ignore any climate targets in our consideration, the emission of greenhouse gases would still be scarce, as there are only limited deposits of fossil energy carriers on earth and their extraction is associated with costs (oil rigs, lignite excavators, etc.). However, discussing the possibility that oil, gas or coal could run out in the long term, a concern that was quite widespread in the 20th century, is largely superfluous given today's state of knowledge and technology. The associated global warming would pose much more serious problems for us much earlier than the scarcity of oil, gas and coal.

The instrument of emissions trading allows an entity, such as a company, within the EU Emissions Trading System to have the freedom to choose to emit or not to emit GHGs, making the scarcity relative. The decision to emit or not emit GHGs depends on the costs or benefits that arise for the entity from the emission of GHGs. These, in turn, depend on the total supply and demand, i.e., the remaining emissions budget and the total need for GHGs of all entities.

The EU Emissions Trading System is predicated on the fact that the number of certificates issued annually is set to align with climate targets. If the number of certificates decreases, GHGs become scarcer and the price for an emitted ton of CO_2-eq increases. This—the concept of emissions trading—will encourage entities to switch to CO_2-neutral or CO_2-low technologies. The fact that this does not always work has been demonstrated by the EU Emissions Trading System since its inception in 2005. The reasons for this are varied and require a more detailed investigation. A primary reason for the ineffectiveness of the EU Emissions Trading System in the first 15 years, however, was the excessive number of emission certificates issued annually, which often surpassed the amount of GHGs emitted and caused the price to fall (Umweltbundesamt, 2022).

We specifically use the term 'entity' here, as an emissions trading system is conceivable not only for companies and their industrial installations, as in the case of the EU Emissions Trading System, but can also be applied to states, industries, or individual citizens.

In summary, within the framework of emissions trading, GHGs are classified as a relatively scarce resource.

14.8 Greenhouse Gas Emissions as an Absolutely Scarce Good

Alternatively, one could argue that GHGs should not be considered as relatively scarce, but as absolutely scarce goods. What might be the reasons for such a viewpoint? As previously described, a good becomes absolutely scarce when the freedom of choice between this and other goods is lost. Therefore, we need to consider whether there are instances in which GHGs are not substitutable, neither on the supply side nor on the demand side. Let us first consider the supply of the good: the remaining budget of 1024 gigatons of CO_2-eq in 2023. If we accept the 2 °C target as a premise, this budget (the supply of the good) is at least fixed in the short term. In the long term, an increase in the budget is conceivable, by removing CO_2 from the atmosphere on a large scale, through reforestation, alternative agricultural techniques, geoengineering such as Direct Air Capturing (DAC)[6], or the natural, long-term degradation of GHGs in the atmosphere.

[6] DAC is a frequently discussed technology that is already considered in many long-term climate scenarios. With DAC, CO_2 is directly removed from the air. If used on a large scale, this technology could contribute to climate goals. However, DAC is also subject to the laws of physics (see Chap. 9 *Thermodynamics*) and requires a significant amount of energy to remove CO_2 from

On the demand side, it is not a given that actors always have a choice between emitting GHGs and foregoing them. If we were to decide today in Germany to stop using oil and gas immediately, one consequence would be that millions of people would not be able to heat their homes in winter. Of course, buildings could be heated without emitting GHGs through extensive renovations and energy from renewable sources, but for various reasons (lack of skilled workers, delays in electricity grid expansion, lack of political will, etc.) this would take years to decades. At least in the short term, many people would freeze at home in winter, get sick and possibly even die. The satisfaction of their existential needs would be severely restricted, making GHGs an absolutely scarce good for them. Similar examples can be found for different population groups in Germany and globally, for whom GHGs are an absolutely scarce good.

In summary, it can be said that in the case of the GHGs budget, no clear distinction can be made as to whether the problem is a case of *relative* or *absolute scarcity*. Depending on various factors such as the actors involved, local conditions, the time horizon (see Chap. 8 *Three Concepts of Time* and Chap. 15 *Stocks*), or the view of nature and humanity of the observer, many environmental problems can represent both a *relative* and an *absolute scarcity*.

The respective answer to the question of *absolute* or *relative scarcity* of an environmental problem has far-reaching consequences for societal handling of the problem, which we will discuss in the following section.

14.9 Consequences of Scarcity Classifications in Environmental Problems

From the example of the remaining GHG budget, we have seen that environmental problems can be interpreted as issues of *relative scarcity* as well as *absolute scarcity*. The type of scarcity interpretation that prevails in a society has significant implications for how problems are addressed and thus for the actors affected.

Consider, for instance, the potential implications of a nationwide emission trading scheme that affects all citizens, thus a relative understanding of GHG-scarcity. Suppose all citizens were allocated a certain amount of emission rights. This approach could efficiently reduce the 2023 per capita emissions in Germany, which are currently around eight tons of CO_2-eq. Emission rights could be traded on the market. Individuals whose emissions exceed their allocated rights would need to purchase additional emission rights. Those who emit less could even profit by selling their rights. The number of emission rights available annually could be progressively reduced to meet climate targets.

---

the air, which we may not necessarily have available or may need elsewhere. Furthermore, DAC has not yet been tested on a large industrial scale, and the removed CO_2 must be stored in some form permanently, without any leaks. Therefore, there are still many uncertainties surrounding this process.

However, such a market-based policy, as efficient as it may be according to mainstream economic criteria, would raise questions regarding its

- justice,
- political feasibility, and
- technical feasibility.

In summary, the form of scarcity that dominates societal discourse can be crucial for addressing environmental problems. This, in turn, has implications for the people affected by the resulting measures, such as a CO_2 price.

As outlined at the beginning of this chapter, understanding scarcity is fundamental to comprehending why certain conditions and changes in our environment become problems for us humans; in other words, why environmental problems exist. The scarcity of goods and human needs, both fundamental and imaginary, provide answers to this. We can ask:

- Why is a certain good scarce?
- Is the good necessary for the fulfilment of existential or imaginary needs?
- Is the good substitutable or not (on the supply and/or demand side)?

Answering these questions can facilitate a better understanding of an environmental problem under consideration. The societal handling of the problem can be classified—perhaps the personal assessment of the kind of scarcity of a good is in disagreement with the widespread societal view—and thus, solutions for the problem can ultimately be developed. The usefulness of the concept of *absolute* and *relative scarcity* for developing solutions will be explicitly addressed in the following section.

14.10 Exploring Solution Spaces Through the Concept of Scarcity

It would significantly simplify the tasks of students, scientists, policy advisors, administrative professionals, journalists, and others if there were a single method, a specific tool, or a master plan for solving environmental problems. However, this is rarely the case in practice. At this point, we cannot provide an example of a singular solution—as we do not know any—but only for the opposite. The example of the GHG budget discussed above demonstrates that there are many possible solutions, ranging from CO_2 taxes and certificate trading, to rationing and bans, to voluntary changes in consumption habits and technical innovations. The question that those dealing with environmental problems often ask is: Which solution is a good one, or perhaps even, considering the circumstances and existing knowledge, the best?

By examining environmental problems through the lens of *relative* and *absolute scarcity*, we can begin to narrow down the answer to this question. Depending on whether the scarcity of a good is *relative* or *absolute*, or perhaps both, different tools for a solution may be considered. For instance, if the greenhouse gas problem is a problem of *relative scarcity,* then it may be effective to introduce economic tools such as a CO_2 tax or a trading system for CO_2 certificates. If we conclude that an environmental problem is due to *absolute scarcity*, then other solutions are needed. These would need to consider both the absolute limitation of the good and its non-substitutability on the demand side. Perhaps the climate problem could be better or additionally solved by a strict rationing of GHG or by promoting a more sustainable lifestyle in wealthy consumer societies.

Clearly, understanding the two forms of scarcity does not always provide a straightforward selection of solutions. Often, it is not easy to distinguish whether a scarcity is relative or absolute, which in turn leads to mixed solutions. The climate problem, for example, seems to be solvable neither with a pure approach of mainstream economics *(relative scarcity),* nor solely through bans, rationing and stimulation of a sustainable lifestyle *(absolute scarcity).* In many cases, a combination of the two perspectives may be effective in finding tools that solve the respective environmental problem most efficiently and fairly.

Even if understanding scarcity does not necessarily lead to a master plan for environmental problems, it still helps, as we have demonstrated, to narrow down solution spaces and to approach the best solutions for environmental problems. We therefore encourage those working on environmental problems to apply the scarcity lens in future considerations of environmental problems, to experiment with both the relative and the absolute perspective, and to decide which of the two is more suitable for recognizing the observed environmental problem most accurately.

Abbreviations:

GHG	Greenhouse Gas Emissions
CO_2-eq	Carbon Dioxide Equivalents

References[7]

Baumgärtner, S., Becker, C., Faber, M., & Manstetten, R. (2006). Relative and absolute scarcity of nature. Assessing the roles of economics and ecology for biodiversity conservation. *Ecological economics, 59*(4), 487–498.

Faber, M., & Manstetten, R. (2010). *Philosophical Basics of Ecology and Economy.* Routledge.

Umweltbundesamt. (2022). *Der Europäische Emissionshandel.* https://www.umweltbundesamt.de/daten/klima/der-europaeische-emissionshandel#teilnehmer-prinzip-und-umsetzung-des-europaischen-emissionshandels.

[7] The contents of this concept are based on: Baumgärtner et al. (2006) and on Faber, M., Frick, M., Zahrnt, D. (2019) MINE Website, Absolute & Relative Scarcity, www.nature-economy.com.

Stocks: Transformation Takes Time

15

Contents

What's it About?

This chapter is about understanding the dynamics of nature, economy, and society. The concept of stocks presented here provides a productive perspective on these relationships. The stock perspective illustrates how things change over time and why some things react very slowly to change, i.e., are consistent.

What initially sounds like a description has significant practical implications: Only by understanding what changes, how it changes, and especially how quickly it changes, can we understand what options for action arise. Additionally, an analysis of stocks provides orientation regarding the question of how much time we need to plan for certain changes. This also helps us understand where we should act as quickly as possible to initiate the necessary changes in time.

M. Faber et al., *Sustainable Action in Economy and Society*,
https://doi.org/10.1007/978-3-662-69122-9_15

15.1 The Stock Perspective

Many people are familiar with the term stock from everyday German language.[1] We use the term *stock (Bestand)*, for example, when we refer to the *inventory* (*Bestandsaufnahme*) of certain products. Or we conduct an *inventory* when we want to get an overview of where we stand and what resources are available to us for a certain task. At the same time, we usually take into account the temporal dimension, speaking of the *persistence (Beständigkeit)* of things, which are characterized by the fact that they do not change or change only slowly over time. This can be meant positively or negatively: If an organization is characterized by *persistence* it can be understood to mean that it is *reliable* even in turbulent times, provides orientation and security. Or it can mean that the organization is incapable of keeping up with the times and does not carry out the necessary developments as it should. But what does the term stock, which we know from everyday German language, have to do with the question of the interplay of nature, economy, and society? How can it help us in the analysis of environmental problems?

The reason we first presented the everyday German meaning of the term is to highlight the analytical potential of the term, which we want to use for dealing with environmental issues. Both in everyday questions and in the analysis of environmental problems, the concept of stocks initially helps to identify *what is*. In addition, with a perspective of stocks, the investigation of the *change of this stock over time* is closely linked to the *inventory*. So if we clarify, for example, how much of a certain product we have in stock right now, such as books in a bookstore, we can also directly ask how this quantity will change. Where will the stock be at the end of the month? How many of the books will be sold, and what new deliveries can be expected? Here it becomes clear that the term stock first looks at the status quo at a certain point in time and in the next step directly enables us to analyze the change of this status quo over time.

This perspective is of central importance in the analysis of environmental issues. At the center of the sustainability debate is the question of what we need for the long-term life and survival of humanity. If we want to formulate it more ambitiously, the question is: "What must be available to future generations to give them the chance to lead a good life on earth?"

Anyone looking for an answer to this will first mention the things that are essential for life and survival: sufficient water, air and food, building materials for the construction of houses, energy sources to heat these houses or enable production. All of these are material foundations of our life—which in many cases are directly "produced" by nature and made available to us. With the help of the stock perspective, we can summarize these things under the term *material stocks*. But

---

[1] This chapter owes its structure and its central arguments to the second part of the book "Sustainability and the Art of Long-Term Thinking" (Klauer et al., 2013). We thank the authors for permission to use their central insights for this book.

Table 15.1 Differentiation between material and immaterial stocks

Material Stocks	Immaterial Stocks
The term *material stocks* refers to the quantity of tangible items of the same kind that can be registered at a certain point in time. Here are some examples: • Stock of wood in a carpentry warehouse, • Stock of available apartments in a residential area, • Stock of available rare earths within a national economy, • Stock of plastic parts and plastic particles in the sea, • Stock of CO_2 in the atmosphere. People can also be registered as *material stock*, for example: • Qualified employees of a company, • Residents of a city, • Available skilled workers within a national economy.	The term *immaterial stocks* refers to non-material items of the same kind that have a certain durability. Here are some examples: • Stock of laws regulating the handling of industrial waste, • Stock of behavioural patterns and conventions that lead to a reduction in the waste of drinking water in a country, • Stock of available knowledge for building a city's energy supply with renewable energies, • Subsidies for lignite mining.

other, non-material things are also essential for us. For our life and well-being, it is crucial how our coexistence is organized. For this purpose, the institutions of law, politics, economy, science, culture, and religion exist. For the complex interaction in our societies to work, contracts must be adhered to, good laws enacted and enforced, knowledge generated and passed on. The totality of social institutions that ensure this, such as constitutions, economic systems, laws, traditions, conventions, rules, and institutions, we refer to as *immaterial stocks* (see Klauer et al., 2013).

Until now, we have only considered stocks that are necessary or desirable for human life. However, there are also material and immaterial stocks that pose challenges for us: CO_2 and other climate-damaging gases in the atmosphere, waste that we must manage, such as plastic in the sea. On the level of immaterial stocks, we have regulations, processes, and behavioural patterns that favour, for instance, fossil fuel production processes over resource-conserving approaches (see Table 15.1).

In addition to determining the status quo of these stocks, the stock perspective considers how the respective stock develops over time, whether and to what extent it increases or decreases. When dealing with sustainability issues, it is crucial to know whether vital material and immaterial stocks are available and what changes can be expected over time. Equally important is understanding how material pollutant stocks and problematic immaterial stocks, such as environmentally harmful behavioural patterns, are maintained and developed, and how we can manage them.

Important to Know: What Does it Mean to Adopt a "Stock Perspective"?

- The stock perspective focuses on *temporal sequences* in the investigation.
- How does a certain material or immaterial stock develop/change?
- What are the lifetimes and rhythms of change of a stock?

15.2 The Natural Basis of Life: The Need for Persistence

The sufficient availability of natural resources and the avoidance of pollutants are essential for a good life on Earth today and in the future. Therefore, for the sustainable development of our societies, it is crucial to monitor the changes in stocks of natural resources and pollutants. The extraction of resources from the environment must be regulated in such a way that the stocks can recover over time. Similarly, when disposing of waste into the environment, we must ensure not to exceed its absorption capacities. Nature serves as a vital service provider for us, ensuring, with appropriate treatment, that essential natural resources and absorption capacities are available in sufficient quantities. However, to maintain this, it is important to understand how much of the natural resources we can "take from the stock" without risking overuse to the point where they can no longer regenerate. Humans share the Earth's habitat with other living beings, who contribute to our natural basis of life and partly become the basis of life for us. In technical terms, this is referred to as a *biocoenosis, a natural community of life,* and particularly refers to a network of interrelationships that connect the different living beings in a specific habitat, such as a forest area or a lake with its shore landscape, and in which they are of importance to each other.

If we consider the entire planet as a community of life, it is quite astonishing that the Earth, throughout its history, despite significant changes and developments, has always been a place that could provide sufficient natural basis to ensure life. Will this fact remain guaranteed in the face of the changes caused by humans on Earth? Like other living beings, the fulfilment of human needs takes priority. But unlike other living beings, humans, as rationally planning beings, have managed to subjugate nature and other living beings to human goals. More than any other living being, humans are capable of fundamentally influencing their environment. And although other living beings are capable of "overusing" natural resources, overuse by humans has a far greater impact and can ultimately become a threat to humans themselves. The stability of the interactions between humans and the natural basis of life depends on moderate usage.

Among the immaterial stocks, those that maintain the ability of a society to act are particularly important. Here, the constitution is particularly worth mentioning, which ideally provides the framework within which disputes and dialogues about sustainability and justice can be conducted, in such a way that successful political action emerges from them.

15.3 Environmentally Harmful Production and Behaviour: Where Persistence is a Problem

Persistence becomes a problem, firstly, when stocks that damage or endanger the natural and social foundations of human life persist or even increase quantitatively. Here, the current situation must first be analysed and then the question of the changeability of these circumstances must be asked: How quickly can we, for example, reduce the stock of waste by recycling existing waste and ensuring that less new waste is produced? How quickly can we develop and enforce new rules that promote resource-saving economic practices? In what way and how quickly can we ensure that long-practiced environmentally harmful behaviours are replaced by environmentally friendly behaviour patterns? Both for material and immaterial stocks, a crucial question for the implementation of changes is how and in what time frame it can succeed to "get the stocks moving." The gravity of these questions is illustrated by the climate problem. Even if humans were to stop emitting any more climate-damaging gases into the atmosphere, the gases already present in the atmosphere and their effects must be taken into account for a long time. In terms of their natural reduction, they are very persistent, as it takes thousands of years for CO_2 emitted by humans to be completely degraded in the atmosphere by natural physical and biogeochemical processes (Federal Environment Agency, 2022). This shows the inertia of natural stocks. However, the climate problem is mainly due to an inertia that lies in human behaviour. Every year, large amounts of CO_2 and other greenhouse gases are released into the atmosphere worldwide from production and consumption. The behavioural changes required to achieve climate neutrality cannot be brought about overnight, as both political processes, such as changes in laws, and changes in production processes and consumption patterns take time.

Persistence is also a problem when stocks, whose preservation is necessary or desirable, are threatened in their persistence. For example, there are many stocks in nature that reproduce themselves over a long period of time, e.g. tropical rainforests or the diversity of animal and plant species. Humans seemingly do not have to do anything for these stocks because they are there by themselves and renew themselves. But over time, such stocks decrease first imperceptibly, then increasingly noticeably in number and quality. The cause is human behaviour, which, without paying attention to the danger, or accepting it, damages and destroys the habitats of species and ecosystems. The continued existence of important stocks in nature is no longer a given, but requires far-reaching behavioural changes and decisive societal and political action (see Chap. 6 *Responsibility*).

Understanding the Stock Perspective

- The stock perspective facilitates an understanding of the dynamic interplay between nature, economy, and society. It highlights areas of inertia and identifies when and where opportunities arise to initiate changes.

- In terms of individual expectations, the stock perspective helps to manage frustrations. It demonstrates that a great deal of patience is often required by highlighting the periods within which change can occur (see Chap. 7 *Power of Judgement*). This is not to justify the slowing down of necessary changes by interest groups, for instance. Instead, the stock perspective should assist in recognizing the foresight required to implement changes effectively and timely (see the concept of *inherent time* in Chap. 8 *Three Dimensions of Time*).

To illustrate the usefulness of the stock perspective in analyzing environmental problems, we will use the example of the Emscher River cleanup, which we previously discussed in Chap. 6 *Responsibility*. However, before we delve into the analysis, here are the key facts:

The Emscher is a river in North Rhine-Westphalia that traverses large parts of the Ruhr area from its source near Dortmund to its mouth in the Rhine near Dinslaken. Spanning approximately 100 km, it flows through Germany's most historically significant industrial region. The ecological problems associated with industrial production are particularly evident in this region, as demonstrated by the Emscher, which transformed from a natural, clean watercourse to arguably the dirtiest river in Germany during industrialization.

The exploitation and pollution of the river began around 1800, and as industrial production in the Ruhr area intensified, so did the burden on the river. The development of the Ruhr area as a mining and industrial region from 1850 led to the construction of massive production facilities, underground coal mines, aboveground ironworks, and transport routes. Additionally, hundreds of thousands of workers and their families had to be accommodated and provided for. Industrial waste and mining residues mixed with human waste in the region's rivers, leading to the deterioration of entire water systems and the flora and fauna on their banks year after year.

As we detailed in chapter 8 on *Responsibility*, it was not until 1899 that a state institution, the Emscher Cooperative, intervened in the unregulated exploitation and pollution of the Ruhr area's rivers due to the untenable hygienic and health conditions. Following this intervention, it took nearly 200 years from the start of the pollution for a comprehensive cleaning and renaturation program to begin in the 1990s, which concluded in 2020. The fact that such a long time elapsed from the emergence of a seemingly clear and regionally confined problem like river pollution, through the recognition of the need for action, to the resolution of the problem, may seem outrageous. For those concerned about environmental protection, this lengthy period is undoubtedly frustrating. We do not wish to dispute these sentiments in the following explanations. However, we argue that beyond the justified anger about the long untenable conditions, the stock perspective should first be used to analyze the situation. Such an "inventory", we propose, helps to identify how changes can be accelerated. The need for action can be better assessed, and a realistic timeline can be established for the necessary changes.

15.3.1 Recognising the Problem: Mobilising Material and Immaterial Stocks

The issue became apparent with the opening of the coal mines in the Ruhr area from 1850 onwards. The region's rivers, which had primarily been manipulated for agricultural purposes through straightening and drainage, became receptacles for industrial waste, sewage, and other refuse. Unwanted by-products or *joint products* of human life and economic activity in the Ruhr area found their way into the rivers, accumulating into problematic *material stocks*. The original stocks of fresh water, fauna, and flora in the Ruhr area's rivers were destroyed within a few decades, replaced by open sewers devoid of life. This situation, exacerbated by the threat of infectious diseases such as typhus and cholera, made the accumulation of sewage and waste in the Ruhr, Lippe, and Emscher rivers an urgent issue for the cities and municipalities. However, initially, no one took responsibility for the problem due to the lack of rules, laws, or regulations specifying who should seek a solution. Although the cities and municipalities were aware of the problem, they ignored it. Just as waste and sewage had accumulated materially, so had years of political inactivity formed an *immaterial stock* that needed to be addressed. To initiate a solution, it was necessary to define clear responsibility rules and establish institutions to enforce these rules. The problem was finally addressed in 1899 with the establishment of the Emscher Cooperative, which was equipped with the necessary knowledge and executive competencies to develop and implement a solution for the Ruhr area's (waste) water management.

Before the identified material and immaterial stocks could be mobilised and solutions implemented, new immaterial stocks had to be established: knowledge and skills, rules and laws, and functioning political and bureaucratic processes. When the Emscher Cooperative began addressing the problem, material stocks also started moving on a massive scale: it was decided to repurpose the Emscher from an independent river course to the Ruhr area's drainage canal. In exchange, the Ruhr and Lippe rivers were rehabilitated to ensure the region's fresh water supply. The Emscher was lowered by three meters and shortened from its original 109 km length to 83 km. Planning knowledge and skill were utilised to move thousands of tons of soil in a region riddled with mining tunnels and pits. Finally, the river course was embedded in a cement canal over its now 83 km length to increase its flow speed and quickly transport away the sewage.

So, when we discuss the need to mobilise material stocks to establish a functioning sewage management system in the Ruhr area at the end of the 19th century, we mean this quite literally. And only after this enormous initial immaterial and then material effort did improvements occur with regard to the actual issue at hand: the waste and sewage stocks in the Ruhr area's rivers.

The pragmatic decision of the Emscher Cooperative worked technically well throughout the 20th century. However, despite representing an improvement for the entire Ruhr area, it remained a considerable ecological and hygienic burden locally. The Emscher ran through the cities as a foul, toxic sewer. There was no longer any fauna and flora in and around the river, and for people, in addition to the smell, there were significant health risks. High concentrations of heavy metals, as measured in the river, can cause symptoms such as sleep disorders, respiratory diseases, skin diseases, intestinal diseases, and more in people who come into contact with the water. In growing individuals, heavy metals such as lead or chromium can damage the circulatory system and the central nervous system, as well as lead to severe learning difficulties.

15.3.2 The Beginning of the End of the Interim Solution: Changes in Legislation

The health risks associated with the Emscher river system led to the initiation of its cleaning and renaturation in 1991, almost 200 years after human intervention began and nearly 92 years after the founding of the Emscher Cooperative. This complex undertaking required the development of immaterial stocks in the form of well-crafted laws (see Chap. 7 *Power of Judgement*) and effective procedures. The starting point for the renaturation of the Emscher was the European Water Framework Directive of the EU, adopted in 2000. This directive, a binding framework for national legislations, introduced strict limits over time for water pollution within the European Union (see Chap. 8 *Three Dimensions of Time*). If the condition of the Emscher had remained unchanged, it would have been evaluated by Brussels as a blatant violation of the Water Framework Directive by 2015 at the latest, potentially resulting in high fines. This legislation compelled the politically responsible to act. After nearly 200 years of accepting the unfortunate state of the Emscher as a pragmatic solution, it was time to take *political responsibility*. The creation of such a situation by a law can be seen in retrospect as evidence that the authors of the European Water Framework Directive acted with power of judgement.

Just as the founding of the Emscher Cooperative led to long-term effective changes in material and immaterial stocks, the announcement of the base year 2015 for mandatory compliance with European water quality limits initiated a process of fundamental change. This process involved step-by-step changes in immaterial and then material stocks. The *Blue Emscher* project was launched, pooling knowledge, skills, and financial resources to initiate Europe's largest water infrastructure project. An underground sewerage system made of 35,000 pipe segments was built, running through 12 cities at a depth of 30 m. A sewage system was installed at over 260 individual construction sites and connected to four newly built sewage treatment plants. More than 4.4 billion euros were mobilized for this massive undertaking. These figures clearly illustrate the amount of material moved, money spent, people mobilized, knowledge pooled, and decisions made to solve a regional problem.

In 2009, the Emscher Cooperative decided to go beyond cleaning the Emscher river water and also tackle the renaturation of the entire 350 km long river system, including the tributaries and branches of the Emscher. This decision was possible because heavy industry, which had been a significant part of the Ruhr area's economy for almost two centuries and had significantly shaped its history, has largely withdrawn from the Ruhr area and relocated to other regions of the world. This change brought enormous economic and social challenges for the cities and municipalities of this region, but it greatly reduced the environmental burden.

There is hope that the stocks of plant and animal life in the river and on its banks will continue to recover and a functioning ecosystem will form. Despite the mobilization of impressive immaterial and material stocks, humans can only partially influence and slightly accelerate this development.

Summary

The stock perspective provides a method to analyze the necessary conditions for changes. The following questions must be asked:

- Which material and immaterial stocks are important for understanding the problem to be addressed?
- How do these stocks manifest the status quo?
- What dynamics need to be initiated to achieve an intended change?
- What inherent times/own times do these stocks have?
- What do these own times mean for the overall timeframe that must be set for the implementation of the intended change?
- Can conditions be formulated that help to recognize the *Kairos* (see Chap. 8 *Three Dimensions of Time*), i.e. the right time window for certain steps?

References[2]

Klauer, B., Manstetten, R., Petersen, T., & Schiller, J. (2017). *Sustainability and the Art of Long-Term Thinking*. Routledge.

Umweltbundesamt. (2022). *Die Treibhausgase* [Text]. Umweltbundesamt. https://www.umweltbundesamt.de/themen/klima-energie/klimaschutz-energiepolitik-in-deutschland/treibhausgas-emissionen/die-treibhausgase.

Reading Recommendations for Further Reading on Part 4 "The Interplay of Man and Nature"

Daoud, A. (2018). Unifying studies of scarcity, abundance, and sufficiency. *Ecological Economics, 147*, 208–217. [Scarcity]

Dasgupta, P. (2021). *The economics of biodiversity: The Dasgupta review*. Hm Treasury. [Stocks]

[2] The contents of this concept are based on: Faber, M., Frick, M., Zahrnt, D. (2019) MINE Website, Basics of Life—Stocks, Stores & Funds, www.nature-economy.com.

Dyckhoff, H. (2023). Proper modelling of industrial production systems with unintended outputs: A different perspective. *Journal of Productivity Analysis, 59*(2), 173–188. [Joint Production]
Funtowicz, S. O., & Ravetz, J. R. (1990). *Uncertainty and quality in science for policy*. Springer Science & Business Media. [Ignorance]
Gleick, J. (1988). *Chaos: Making a New Science*. Heinemann. [Ignorance]
Norgaard, R. B. (1990). Economic indicators of resource scarcity: A critical essay. *Journal of Environmental Economics and Management, 19*(1), 19–25. [Scarcity]
Raina, R. S., & Dey, D. (2020). How we know biodiversity: Institutions and knowledge-policy relationships. *Sustainability science, 15*(3), 975–984. [Stocks]
Shackle, G. L. S. (1955). *Uncertainty in Economics*. Cambridge University Press. [Ignorance]
Smithson, M. (1988). *Ignorance and Uncertainty: Emerging Paradigms*. Springer. [Ignorance]
Taleb, N. N. (2010). *The black swan: The impact of the highly improbable*. Random House Trade. [Ignorance]

Part V
The Unity and Incompatibility of Man and Nature

Foundations of Life, Orders of the Living, and Orientations of Humans

16

Contents

16.1 The Remarkable Sustainability of Life

Before the appearance of humans, life on Earth demonstrated its sustainability without the need for rational planning. Despite significant disruptions, such as meteorite impacts associated with mass extinctions, life reproduced in a manner that allowed a variety of increasingly complex forms to emerge throughout its evolution (see Chap. 11 *Evolution*). Its foundations were consistently preserved or renewed over several billion years.

Among the earliest living organisms, there were those that, like the single-celled, nucleus-free blue-green algae (prokaryotes), have played a crucial role in the preservation and development of all life from then until today. They released free oxygen into the Earth's primordial atmosphere, where oxygen did not exist in free form, as a by-product of their metabolism (see Chap. 13 *Joint production*),

M. Faber et al., *Sustainable Action in Economy and Society*,
https://doi.org/10.1007/978-3-662-69122-9_16

which gradually accumulated to a significant level of oxygen in the atmosphere (see Chap. 15 *Stocks*). This was and still is the prerequisite for the emergence and continued evolution of organisms with *aerobic* (i.e., oxygen-mediated) respiration which have shaped the Earth up to the present day. Oxygen respiration is typical for all complex multicellular organisms *(Eukaryotes)*. Today, in addition to blue-green algae, plants contribute to the maintenance of the oxygen atmosphere through their photosynthesis.

While a single bacterium can be distinguished from its kind and its environment under a microscope, evolution has gradually produced more extensive and internally differentiated living beings. These organisms, classified as plant and animal organisms, are clearly distinct from each other and strive to maintain their unique existence. They live not only for others but also for themselves. They come into being, grow, maintain themselves in their developed form, reproduce, suffer, and die. Animal organisms show the tendency to preserve their life if possible, to avoid harmful influences, and to flee from threats. Another tendency of such organisms is to pass on their life to the next generation through reproduction. A third tendency, which we already noted in the blue-green algae, remains inevitably imprinted on the life of all organisms by their environment: With their life, often also with their dying and beyond death, they contribute to the persistence of the respective ecosystem[1] , insofar as they form part of its basis for life.[2]

Key Point: The Interconnection of Life
Organisms do not live solely for themselves, but contribute to the continuity of their ecosystem and to the survival of life as a whole.

16.2 The Three Tele

Individual organisms, species, and biological communities must always be understood in their mutual dependencies. Accordingly, one can consider an organism in three respects: 1) In its relationship to itself, i.e., in its individuality or uniqueness, 2) in its relationship to its reproductive community (population or species), and 3) in its relationship to a larger community of organisms, such as a biological community or the community of all living beings on Earth. This results in three fields of relation for an organism. Within these fields, the existence and development of an organism are determined by three tendencies, which in the following we will

[1] An ecosystem is a community of organisms of several species in a certain space, such as a lake with its shoreline or a forest area. It therefore consists of living and non-living components.

[2] A detailed presentation of the relationships presented in this chapter can be found in Faber and Manstetten (2003, Chap. 10–12).

refer to as *Tele*. *Tele* is the plural of the Greek word *Telos*. In Greek, it means the purpose, goal, end, or outcome of a matter, in a broader sense also the order that is decisive for the temporal development of a thing and its conclusion (see Chap. 8 *Time*). In our context, Telos can be understood as determination, determining order, or determining orientation. We will subsequently speak of the three Tele of an organism (Faber & Manstetten, 2003 [2010], pp. 116–136 [pp. 84-96]; Spaemann & Löw, 1991).

Important to Know: The Three Tele

1. Self-preservation and self-development: This is the Telos by which an organism relates to and organizes itself. Through this Telos, the individual, distinct organism is constituted.
2. Self-repetition or self-renewal through reproduction: This is the Telos in which an organism relates to its population or species and contributes to their self-organization. Through this Telos, the species is constituted.
3. Serving, the alienation and self-alienation in relation to other organisms: This is the Telos by which an organism relates to organisms of other species. Through its life and also through its death, an organism becomes part of the life basis of other species, contributes to their development and to the self-organization of its biological community. Through this Telos, the ecosystem is constituted.

These three Tele do not have to be conscious to a particular organism; it realizes them simply through its mere existence. The third Telos plays a special role. From its perspective, the life of the individual points beyond the preservation of itself and its species. This serving can take the form of a peaceful relationship between different species (symbiosis[3], commensalism[4]). For example, flowering plants serve bees and other insects by providing nectar. The insects transfer pollen, thus enabling the flowering plants to realize the second Telos, reproduction. The grass gives its leaves, which provide food for grazing animals. However, often the realization of its third Telos for an organism ends the possibility of continuing to realize its first and second Telos. For the antelopes in the wild to serve lions as a basis for life, they must suffer a painful death at the end of a hunt. But the services of the antelopes are not one-sided, because the hunting animals also serve—both the ecosystem as a whole and the species that depend on this system. Predators ensure that the populations of prey animals do not get out of hand. In addition, they serve scavengers who feed on the remains of the killed animals. Furthermore, the predators themselves become the basis of life for countless microorganisms with their

---

[3] Symbiosis refers to an interaction between individuals of two species that is beneficial to both.

[4] Commensalism refers to an interaction between individuals of two species that is advantageous for members of one species and has no effect on the other.

excrements, and finally also with their corpses. If such service recipients, i.e., the predators, are missing, this can result in fundamental disturbances in the ecosystem. This is shown by the history of rabbits in Australia, which were not perceived as prey by any animal species and became a major problem not only for humans but also for the local flora and fauna. In general, species that reproduce strongly can get out of hand if there are no natural predators in the corresponding ecosystem. Among other things, with the argument that the natural equilibrium in near-natural forests must be restored, the presence of wildcats, lynxes, and wolves or the immigration of brown bears is currently being promoted in Central Europe.

The realization of the three Tele in an ecosystem and thus the shape of the ecosystem itself is changeable over time. When a new species appears in the course of evolution (see Chap. 11 *Evolution*), it can open up new services for its self-preservation and self-development, but can also itself be perceived as a giver of new services.

16.3 The Third Telos and the Foundations of Life—The Concept of Funds

In the previous chapter *Stocks*, we mentioned that some stocks serve as reserves for living beings. Now, we want to ask: Where do these reserves come from? To answer this, it is useful to consider the services provided by living beings that directly or indirectly support the life of other living beings. We refer to these supports as services, and to the living beings that receive them, as service recipients. These services can be provided by both living beings (e.g., animals or plants) and non-living sources (e.g., the sun or capital goods such as houses and machines). We refer to a source of services for one or more types of living beings as a *fund*. The term 'fund' originates from the Latin word *fundus*, meaning 'land and soil', but it can also refer to the essential foundations of a matter. Using the pair of terms 'fund'/'service', the question of the basis of life can be formulated as follows: Which funds and services are necessary for the existence of a biological community, a species, or an individual living being?[5]

Certain primary funds are essential for all species. For example, the sun provides services in the form of energy (heat and light), and the earth's water cycle, including seas, clouds, watercourses, lakes, and groundwater, is an indispensable basis for all life. Other primary funds, important for many living beings, can be summarized under the terms 'soil' and 'air'. However, most funds depend on the services that species provide or use within a certain community. Beyond the habitat, the biotope, animal species especially need plants or other animals for

[5]The considerations presented below on a theory of funds owe significant impulses to Chap. 9 from Georgescu-Roegen's groundbreaking book "The Entropy Law and the Economic Process" (1971).

survival. Their funds are therefore species themselves. These usually belong to the same ecosystem as the service-receiving species. Some services of funds are needed in such a way that they leave the life of the living beings contained in them untouched and unharmed. Such a service is, for example, the shade provided by trees. But often the services are specific parts of living beings, such as the fruits, flowers, and leaves of a plant, which are consumed by the service-receiving species. In some cases, the service is a whole living being itself, when it serves as food for another living being and thereby dies.

The time structure of funds should also be taken into account (see Chap. 8 *Three Dimensions of Time*). Apart from primary funds like the sun, they do not have a permanent substrate, since they belong to life themselves. Their continuity results from the constant renewal of the elements of the fund. This presupposes the first and second Telos of living beings, i.e., the Tele through which a species, mediated by the individuals, maintains and renews itself. The first Telos involves a constant throughput of matter and energy (see Chap. 9 *Thermodynamics*), the second Telos with death and birth, the disappearance of individuals, and the emergence of new ones in their place. In the coming and going of individuals, an ecosystem has its persistence, i.e., its funds can remain stable over a long period of time despite changing composition.

Important to Know: Funds and Services

A fund is a source of services for one or more types of living beings. Plant and animal species within an ecosystem can be both funds and service recipients. As service recipients, they appear from the perspective of the first and second Telos. From the perspective of the third Telos, they are seen as funds. For the persistence of a fund, the first and second Telos of the living beings involved are necessary prerequisites. By studying funds in their temporal development, important insights into the sustainability of of ecosystems can be gained.

The equilibrium within a biological community is achieved through the interplay of all species within a network of services. If essential services are removed from such a community, there can be an explosive growth of certain species, or species lacking the necessary resources may perish; both scenarios can also occur simultaneously. Continuous research is necessary to identify the diversity of funds and the corresponding quality of services within ecosystems. For instance, trees are funds in many respects. Not only do their flowers and fruits serve as funds, but also their leaves, bark, and the shade they provide, are services for a multitude of creatures. After their death, their deadwood also offers further services, provided humans do not intervene. The same applies to the soil, with its surface in some places already containing an innumerable number of microorganisms in a single lump of clay. In particular, those services of funds that return everything that might appear as waste from a human perspective back into natural cycles should be emphasized. A large number of different bacteria and fungi play a particularly important role in natural disposal. Disposal in nature can be regarded as recycling in the sense of a complete material cycle. The amount of biomass produced over a period of time, i.e., the mass that becomes part of living organisms, must correspond to an approximately equal amount of material to be "disposed of." Ultimately, everything that has once been produced in nature enters a cycle in which it is taken up as a service. All decay and decomposition processes are at the same time life processes of living beings. In nature, therefore, production, consumption, and disposal cannot be separated; this separation only applies to the human economy.

Another characteristic of funds in nature is joint production (see Chap. 13 *Joint production*). As a rule, funds produce more than one service at the same time. A tree, in realizing its first two Tele, necessarily produces leaves and fruits, bark and shade as services for other living beings. A very important service, which is produced by all green plants as a joint product of their photosynthesis, is the oxygen that other higher living beings need for breathing. Some natural communities, such as tropical rainforests, 'produce' their own climate in the interplay of their funds.

All these statements no longer apply when a dominant species appears in biological communities, which can make other species subservient and exploit them at will, while largely exempting itself from the orders of service. If such a species spreads globally, it will endanger the lives of other species and thus also its own, and in the extreme case, it will put the evolution of life itself at stake. Are humans to be regarded as such a species?

Important to Know: Is Nature Without Humans an Ideal State?
Nature should not be glorified. Massive die-offs through disasters such as earthquakes, volcanic eruptions, floods, climate changes, and meteorite impacts is part of evolution. The equilibrium of natural ecosystems includes "eat and be eaten". This means a constant threat for prey animals as well as pain and suffering when they become the prey of their respective predators.

16.4 The Three Tele and Humankind

When we observe an animal growing up in the wild, we can roughly determine its basic structures of existence and its behavioral patterns in advance. This allows us to make certain statements about its self-preservation and self-development, its first Telos: We know what habitat it claims, which animals or plants it will consume, which enemies will be dangerous to it, how it typically moves, how many offspring it will produce under normal circumstances, how old it can become. Humans, too, as living beings, are determined by the three Tele. But something crucial is different in humans. Humans determine themselves, or at least they try to. Humans want to freely choose how they live, preserve, and develop their lives—for themselves and in community with other humans. The idea of individual freedom, which forms the core of self-determination, is the reason why the personal form of self-preservation and self-development of a human—unlike that of a fly, a seal, or a zebra—cannot be easily determined in advance.

The same applies to reproduction, the second Telos in its manifestation through humans. If a pheasant or a cat does not reproduce, it is usually because these beings are prevented from doing so by external circumstances or a physical defect. If, on the other hand, humans remain childless, it may be that they do not want to reproduce by conscious choice and thus oppose the second Telos out of free self-determination. Even more so, they can oppose the third Telos, serving, as far as it is in their power. Isn't it a characteristic of being human that a person resists serving other living beings as a basis for life as much as possible? Although humans cannot completely remove themselves from the three Tele, they have a special relationship to them. We want to address this relationship with the term *interest*.

If life, as far as it can be represented within the three Tele, is intrinsically sustainable, one can ask: What does it mean when the factor of interest, which we define as particularly applicable to humans, enters this area of the Tele? The sustainability of life, which seems self-evident from life itself, can be endangered by humans.

16.5 The Interests of Humans

Due to their self-determination, humans give the three tele different manifestations. In these manifestations, the three Tele are modified and transformed. In extrem cases, they can become unrecognizable or turn into their opposite. The processes of modification and transformation of the Tele are guided by *interest*.

The word *interest* originates from the Latin *interest;* this term means: Something *is of importance*. By recognizing their interest, a persons become aware, imagine or believe that something is important to them. Interests guide actions: Persons decide to act according to what appears important to them. Awareness or imagination belongs to the realm of thinking, while decision-making belongs to the realm of willing. Thinking and willing introduce the element of freedom into every interest—albeit always within wider or narrower scopes.

Instead of *Tele,* one can refer to certain *classes of interests* in relation to humans. In the context of our considerations, the term *interest,* like the term *Telos,* refers to tendencies or orders in which humans find their life and themselves. Unlike *Telos,* interest always expresses a more or less free behaviour, an agreement, reshaping or rejection of given tendencies. Interest contains a certain tendency (i.e. the Telos), but also the possibility to shape it freely or to distance oneself partially or even completely from this tendency. This is why concrete interests can be negotiated. Persons can decide to withdraw their interests or to redefine them in new and different ways.

Important to Know: The Three Classes of Interests

We describe three basic forms in which humans can shape the three Tele with the three classes of interests. Interests can relate to the individual person, to groups and communities, or to all of humanity in its relationship to nature. Accordingly, we distinguish three classes of interests, which will be abbreviated as interests in the following:

1. The first interest arises from the perspective of the individual person: self-interest.
2. The second interest corresponds to the perspective of the community: the community interest.
3. The third interest takes the perspective of sustainability. It is the interest in preserving the natural basis of life and in the good life of humans together with all creatures: the interest of humanity.

16.6 The First Interest—Self-Interest

Self-interest, like the first Telos, refers to the self-preservation, self-development, and self-fulfillment of a person. But the decision on how and with what goals persons believe they should preserve, develop, and fulfill themselves, insofar as circumstances allow, lies with the individual; it is not objectively determined. Self-interest includes the choice of those goals, means, ways, and circumstances that a person considers conducive to realizing their first Telos. However, through this choice, it differs from the Telos, which is an unquestioningly accepted determination by the living being. Humans can ask: What daily food do I want, what profession should I choose? Is the flight to a vacation important for my well-being? However one answers these questions for oneself—most persons see their freedom in being able to choose the life that they believe is good for them. If persons live in conditions where they do not have this freedom due to poverty, hunger, and misery, such conditions are considered inhumane. These people may not be able to assert their primary interest, or they may do so inadequately.

The perspective of self-interest is not long-term, as its time horizon often barely extends beyond the expected lifetime of the respective human individual. Persons may therefore believe they can lead a good life for themselves, even if the consequences of this way of life harm the natural basis of life and thus the lives of future generations. If these consequences will not affect their individual life, it is sufficient for them.

The tendency towards short-term self-interest requires counter-steering if a sustainable economy and society are to be achieved. If, in line with mainstream economics, one assumes that humans as economic beings are guided exclusively by self-interest, a regulating instance like the state must take measures so that even selfish persons develop an interest in preserving the foundations of life (Manstetten, 2000). The easiest way is through the wallet. The levying of taxes on climate gases, wastewater, and waste results in a noticeable cost increase for the disposal of pollutants. This leads to the result that people produce or consume less of it out of self-interest. With this instrument, sustainability seems achievable even if the vast majority of people should be fixated on their selfishly understood self-interest. But the question arises where in a society of exclusively self-interested beings the impulse for the introduction of environmental taxes should come from, if this makes life more expensive for all egoists, while the benefits of these taxes may not flow into their private well-being at all. For a democratically constituted society to adopt political instruments such as environmental taxes, there must be an awareness of the necessity to limit self-interest (see Chap. 5 *Concept of Man*).

16.7 The Second Interest—Community Interest

Like every other type of living being, humanity sustains itself through reproduction. However, humanity realizes its second Telos in a way that has no parallels in the realm of living beings. Over the last millennia, the number of human beings has massively increased, so that instead of a few million, more than eight billion people now inhabit the Earth. However, persons can make their second Telos an interest not only by producing and raising as many offspring as possible, but they can also consciously decide against having any offspring at all. Institutions can also reshape the second Telos of humans, for example, by a state prohibiting a woman from having more than one child, as was the case for a long time in the People's Republic of China.

The horizon for the second Telos in humans is the community interest. Humans are highly dependent on communities. From the individual's point of view, the community interest demands to act not in one's own self-interst, but in the sense of a community to which one belongs or identifies with. The community interest manifests itself in various forms.

Important to Know: Forms of Community Interest

- In most cultures, the most original form of community is found in the family or in family-like structures. Communities that are constituted by the transmission of human life form the framework for the second Telos as realized by humans.
- Almost every person is today part of communities in both, private and professional life, such as clubs, businesses, and organizations in the fields of economy, administration, science, and law. The interests of these communities have a more or less formative effect on the way of life.
- The state, as the community within the borders, institutions, and laws in which a person lives, is a particularly significant form of community that develops its own independent interests.

Every community makes moral demands on the individuals who belong to it or are associated with it: there is an expectation of a willingness to align the pursuit of self-interest with the concerns of the community, or possibly to forgo it entirely in favor of the community. Responsibility (see Chap. 6 *Responsibility*) can usually only be perceived in exchange with other people and thus in relation to a community or society. The skills required for good interpersonal coexistence are usually only learned in communities and can also only be practiced in communities.

However, from an ethical point of view, the community interest is not necessarily good. Communities tend to produce a sence of "we-feeling" by developing certain ideas of their identity. This asserted identity enables them to distinguish themselves from other communities and their members. The own identity can be given a special position or a precedence over other identities. Thus, it often happens that people who belong to different communities encounter each other with disrespect, discrimination, contempt, hatred and hostility. Communities can fight each other, or, as states, even engage in war.

16.8 The Community Interest, the State, and the Problem of Sustainability

Community interests are structurally designed for persistence. Generally, the community is not supposed to collapse and perish with the departure of certain members but rather to persist through the turnover of its members. This orientation towards persistence is often reflected in the community's rules and statutes, which are intended to regulate the interaction among its members. This is particularly applicable to the state and its institutions. The constitution of modern states generally obliges state actors such as politicians to refrain from undertaking anything that would jeopardize the continuity of the state. This implies that a state is designed for sustainability. In light of the knowledge about the human-caused depletion of natural resources, the state, with its power to establish rules, is the appropriate entity to address demands for a comprehensive sustainability policy.

However, it seems that many states, despite this commitment to sustainability, are struggling to secure the internal cohesion of their society. A state that tolerates extreme income and wealth inequality, arbitrarily distributes privileges in the interest of certain groups and disadvantages other groups, neglects necessary infrastructure measures, and overlooks education and health, is unlikely to make significant efforts for sustainable development. When autocratic and dictatorial regimes are more concerned with the self-interests of the rulers and their allies than with the concerns of their population, major transformations can hardly be expected. Major transformations will certainly not occur where states are willing to threaten or engage in war (Manstetten et al., 2021). The fact that currently, states with considerable power and economic strength justify military attacks on neighbouring territories based on their alleged national community interest, prompts the states to which these territories belong to prepare for defense. This readiness for defense can, under certain circumstances, also turn into a readiness for war. The result is that efforts urgently needed for a socio-ecological turnaround are diverted to a significant extent into building military infrastructure and armament efforts.

The result is that efforts, which would be urgently needed for a socio-ecological turnaround, are invested to a significant extent in the construction of military infrastructure and armament efforts. When wars actually break out, the associated consequences almost completely overshadow the task of protecting and preserving natural basis of life.

Important to Know: The State and the Interest of Humanity
A state is obliged to ensure its survival. This results in the task of preserving natural basis of life. However, many states only partially fulfil this task. This applies not only to states with corrupt or dictatorial regimes. Natural basis of life have no territorial boundaries, so a state can often achieve little without cooperation with other states. In addition, the commitment to sustainability is not always appreciated by voters in the election cycle. As long as different states and societies do not agree on a common interest of humanity, it can be expected that special advantages for a state or the groups ruling in it are more important than preserving the natural basis of life and the natural world.

16.9 The Interest of Humanity

Ecosystems untouched by humans, as depicted from the perspective of the three Tele, are now the exception on Earth, and their number continues to decrease. The natural structures of a balance between giving and taking services, as they were found before the appearance of humans, today exist almost only as an idea. At the same time, humans as a biological species can evade the third Telos in an unprecedented manner. The services that humans provide to the existing ecosystems on Earth make them appear only in exceptional cases as a fund for life, but normally as a source of harmful services.

Due to the immense potential danger to the natural basis of life posed by the human species, it bears a special responsibility for life on Earth and its future development. The question is occasionally raised whether it would be desirable to return nature to a state in which the transformations by human interests would completely disappear. This would probably result in the disappearance of the human species. It is conceivable that this could happen, but the idea that humanity is aiming for its own extinction is absurd. The fact that humans as living beings can never completely step out of the three Tele argues against this. Like all other species, humanity has the tendency to preserve itself as a species. From the perspective of the second Telos, there must be an interest of humanity in its continued existence.

From the perspective of the second Telos, however, one can imagine that humanity will achieve a more sustainable life through gentle technologies without direct damage to future generations, but possibly all areas of the natural world that are not apparently necessary for humans could be destroyed. However, this calculation does not hold. Because there is fundamental ignorance (see Chap. 12 *Ignorance*) about whether the natural resources that are destroyed or left to perish in this way might actually prove to be essential for humanity. Moreover, in this idea, humanity as a whole appears like an immense, collective Homo Economicus, concerned solely with maximizing its own needs without regard for its fellow creatures. Other species would only be considered to the extent that is unavoidable.

Such a concept of humanity's interest seamlessly continues with the effort, which began at the beginning of the 17th century and persists to this day, to secure the well-being of humanity through maximum domination and exploitation of nature. It cannot be denied that since then, continuous technical and economic progress has led to an increased life expectancy of humans and a high level of need satisfaction in the richer countries of the world. The proportion of the world's population suffering from hunger has overall decreased in the last fifty years, sufficient nutrition, medical care and education are accessible to a larger number of people worldwide than at other times.

Humanity has embarked on this path of nature domination from a perspective that was and is particularly groundbreaking for economics to this day: The interest of humanity was and is seen in the *improvement of living conditions* of humans. This goal was explicitly formulated by the founder of classical economics, Adam Smith. While Smith emphasized the importance of universal benevolence, which extends to all sentient beings and not just humans, the subsequent period brought a one-sided fixation on the well-being of the human species (Faber & Manstetten, 2014, pp. 106–107). The one-sidedness of the interest of humanity understood in this way is evident in the exclusion of other sentient beings. The animals, to the extent that their numbers and diversity are not decimated by overhunting, overfishing, and habitat destruction, suffer under human dominion through factory farming, mechanized slaughter, and scientific experiments that invade their bodies and lives.

Important to Know: The Interest of Humanity and the Domination of Nature

The domination of nature through technology is fundamental to the way of life in modern societies worldwide. The way it has been practiced over the last centuries is not sustainable. The development and deployment of resource-saving and environmentally friendly technologies on a large scale are indispensable for socio-ecological transformations. However, these transformation processes must not be focused solely on the long-term preservation of the current forms of need satisfaction of members of the human species. This approach entails a fundamental injustice towards the life of non-human species and towards nature itself.

16.10 The Interest of Humanity as a Responsibility for Creation: Ideas from Albert Schweitzer and Hans Jonas

Long before the environmental crisis became apparent, Albert Schweitzer formulated an ethic based on reverence for life. True humanity, as Schweitzer realized, is only possible in the community of all creatures, in connection with all plant and animal life. In Schweitzer's view, the properly understood interest of humanity explicitly includes the third Telos of living beings. Being human means allowing oneself to be freely and consciously dedicated to the further development of life on Earth. Hans Jonas (1903–1993) identified threats to the survival of humanity as early as 1979. The responsibility that people must assume pertains to the existence of humanity: "Act in such a way that the effects of your action are compatible with the permanence of genuine human life on earth. Or negatively expressed: Act in such a way that the effects of your action are not destructive for the future possibility of such life" (Jonas, 1979, p. 36). What Jonas refers to as genuine human life is a life that does not oppose other species and the development of nature, but rather acknowledges an intrinsic right of the natural.

16.11 What Does the Interest of Humanity Imply?

The interest of humanity does not equate to an interest that all people living on Earth or a majority of the eight billion people agree upon. If one wants to know what a person's self-interest is, one can ask them for information. The community interest is reflected in the decisions made in the name of the community. In countries like Switzerland, the interest of the people who belong to this state is determined by referendums. However, there is no procedure to clearly determine what the interest of humanity is and what consequences can be derived from it. What can be said about the interest of humanity under these circumstances?

Immanuel Kant urged every human being to respect humanity in their own person to such an extent that no other goal may be placed above this humanity (Kant, 1974, p. 61). With Kant, we assume that, even if perhaps in an unclear and ambiguous sense, the dimension of humanity is present in the consciousness of every human being. The extent to which clear interests, even concrete instructions for action, can be derived from this will remain the subject of controversies and political disputes. However, it can certainly be said: Where the idea of one humanity is taken seriously, where there is a genuine effort to align action with the interest of this humanity, all private self-interests and all limited community interests are put into perspective. The interest of humanity demands not to take private individual and particular group interests too seriously, to reconsider them, to adjust them in the light of higher principles or to withdraw them completely. Forms of need satisfaction and concepts of well-being, which appear self-evident for individuals or for entire social strata, are put to the test, and at the same time every withdrawal opens up for something new. New forms of human life and coexistence with nature appear on the horizon and what was previously unthinkable becomes possible.

Important to Know: Signs of Change
On August 20, 2018, the first school day after the holidays, 15-year-old student Greta Thunberg staged a protest with a sign reading "Skolstrejk för klimatet" ("School strike for the climate") in front of the Swedish Parliament in Stockholm. This marked a significant milestone in the fight against climate change. Regardless of what has been achieved or not since then, Greta Thunberg's school strike has set a sign that gives hope. There will always be, albeit unpredictably and unforeseeably, impulses like those that have come from individual actions like that of Greta Thunberg. It is possible that such impulses will change societies and states and make a decisive contribution to the upcoming transformations.

References[6]

Faber, M., & Manstetten, R. (2003). *Mensch-Natur-Wissen: Grundlagen der Umweltbildung*. Vandenhoeck & Ruprecht. [English Translation published in 2010: *Philosophical Basics of Ecology and Economy*. Routledge.]

Faber, M., & Manstetten, R. (2014). *Was ist Wirtschaft?: Von der politischen Ökonomie zur ökologischen Ökonomie*. Verlag Karl Alber.

Georgescu-Roegen, N. (1971). *The entropy law and the economic process*. Harvard University Press.

Jonas, H. (1979). *Das Prinzip Verantwortung. Versuch einer Ethik für die technologische Zivilisation*. Suhrkamp. [English Translation published in 1984: *The Imperative of Responsibility: In Search of Ethics for the Technological Age*. Chicago Press].

Kant, I. (1974). *Kritik der praktischen Vernunft. Grundlegung zur Metaphysik der Sitten: Bd. VII* (W. Weischedel, ed.; 2. Auflage), Suhrkamp.

Manstetten, R. (2000). *Das Menschenbild der Ökonomie: Der homo oeconomicus und die Anthropologie bei Adam Smith*. Karl Alber.

Manstetten, R., Kuhlmann, A., Faber, M., & Frick, M. (2021). Grundlagen sozial-ökologischer Transformationen: Gesellschaftsvertrag, Global Governance und die Bedeutung der Zeit. *ZEW-Discussion Paper, 21*(034).

Spaemann, R., & Löw, R. (1991). *Die Frage wozu? Geschichte und Wiederentdeckung des teleologischen Denkens*. Piper.

[6]The contents of this concept are based on: Faber, M., Frick, M., Zahrnt, D. (2019) MINE Website, Absolute & Relative Scarcity, www.nature-economy.com.

Summary and Outlook 17

Contents

Our considerations have brought together 14 concepts that seem particularly important to us for understanding and solving environmental problems. However, no manual of environmental and sustainability issues has been presented here that could provide something like a complete overall picture. Instead, we deal with selected aspects. Although it might seem as if the selection was arbitrary, as one of the selection criteria was the fact that a particular topic was for a while or permanently the subject of the research of the group around Malte Faber. But behind this selection, as it may have randomly arisen in detail, there was always the drive to grasp an environmental problem—be it wastewater, waste, or climate, be it the organization of environmental policy, as comprehensively and thoroughly as possible so that no relevant aspect was overlooked. This resulted in the investigations always being interdisciplinary and the approach, even if it started with a special problem, was aimed at revealing general structures. All the questions and topics dealt with were placed in the horizon of possible practice, and the insights gained have to a large extent also flowed directly into policy advice. Therefore, we are convinced that our concepts, taken together, are in principle suitable for highlighting essential and useful aspects of environmental problems in different ways for practical decisions.

In retrospect, we want to mention some key points that were important to us in the previous considerations.

M. Faber et al., *Sustainable Action in Economy and Society*,
https://doi.org/10.1007/978-3-662-69122-9_17

- We have outlined how the fourteen concepts of our book can be used to explore specific environmental problems.
- A sequence of steps was suggested in which these concepts can be studied and their inherent connection can be recognized.
- It was a particular concern of ours to encourage readers to think independently and to encourage transfer based on our approach. This is especially true for topics such as *thermodynamics*, *joint production*, *power of judgement* and *ignorance*.
- With the choice of topics, we have provided impulses for a catalog of interdisciplinary questions with which environmental problems can be analyzed. Differences must be taken into account: Questions of joint production and scarcity are relevant for every environmental problem. Ignorance is a kind of reminder that always needs to be considered in order to appropriately assess the scope and value of knowledge about environmental problems. Temporal aspects, such as *irreversibility*, *Chronos*, *Kairos*, *sense of time* and *stocks*, have been given the attention that is necessary for environmental problems. *Entropy* and *evolution* form a background that is always present, but does not have to be addressed in every case in concrete investigations of specific environmental problems. The significance of concepts like the latter is mainly that they form a horizon within which concrete problems can be meaningfully embedded and considered.
- Our book aims at practice. Even though it is a theoretical text in form, insofar as theories and approaches from different sciences are taken up and further developed, all the concepts presented here are designed with possible action in mind. For this reason, we have given numerous examples. The recipients are therefore people who want to understand environmental problems comprehensively in order to contribute to necessary changes within their capabilities.
- From the concepts of our book to concrete practice, there is a considerable way to go. Because none of the concepts can be directly translated into action. Intermediate steps must be taken to mediate between theory and practice, which cannot be foreseen in advance. Since the mediation depends on the circumstances and the opportunities that open up in them, it requires *power of judgement*.

That some readers may miss topics and questions in this book that seem important to them could not be avoided. But some things that belong to the matter and are still missing here, we would like to address from our side. Since our actual concern is the transition from theoretical concepts, which reveal the problems in their essential features, to practical action, we would like to finally mention three problem areas to which something should be said in the sense of our intentions, but which have not been adequately dealt with in our previous research. These are

- the importance of the distribution of income and wealth,
- the problem of the media,
- the complex of power.

17.1 The Significance of the Distribution of Income and Wealth

On a global scale, income and wealth are very unevenly distributed. In many states, there are extreme differences between rich and poor, and there are also extreme differences between wealthy countries like the USA or Switzerland and poor countries like Mozambique or Madagascar. Inequality is a challenge within societies as well as from a global perspective, regardless of environmental issues and sustainability concerns. The seemingly limitless scope for action on the part of the wealthy is met with powerlessness, dissatisfaction, resignation, or even outrage on the part of those who are excluded from wealth—and thus from the fruits of what humanity as a whole or their own society achieves. Apart from moral evaluations, it should be noted that inequality results in social tensions that repeatedly lead to war and civil war and trigger migration movements. The unequal distribution of income and wealth also influences power structures. In our considerations on the *Homo Politicus* (see Chap. 5 *Concept of Man*), we emphasized the importance of majority capacity and consensus. But apart from the ability of the rich to influence many people through the media, they can delay or completely prevent impending changes through their financial power and their influence on politics with their lobbying work, if they do not even initiate developments that run directly counter to the goals of justice and sustainability. The differences between rich and poor also become clearly noticeable in dealing with natural resources. Compared to other population groups, the per capita consumption of the environment among the rich is many times higher. At the same time, they can protect themselves much better than the poorer ones against negative environmental influences. Moreover, the elites can organize powerful resistance against changes in environmental laws through media and politics. "The resistance of the elites is an inevitable fact, in today's era (with its transnational billionaires who are richer than entire states) at least as much as during the times of the French Revolution" (Piketty, 2022, p. 25).

If mainly left-wing groups often show little respect for the institutions of law, this is due to the fact that existing property relations in liberal constitutional states are fundamentally protected by constitution and legislation. There are virtually no limits set by existing law to the seemingly limitless increase of income and wealth. Law and legislation offer no leverage for far-reaching redistributions. At the same time, it should be noted that radical redistribution measures can in turn lead to massive tensions and conflicts and thus to social instability.

17.2 The Problem of the Media

Environmental and sustainability problems must gain attention in a society in order for them to be perceived as problems and taken up by the economy and politics. Essential for social attention is the way in which problems are communicated.

At the beginning of this communication are the sciences with their hypotheses, theories, and findings verified in transparent procedures. From there, warnings, alerts, and suggestions for changes can originate. However, what reaches political parties, administrations, business enterprises, cultural organizations, as well as citizens, is only partially dependent on the scientists. Because communication is a matter of media—and therefore also a matter of individuals, companies, and organizations that understand how to handle media communication via different channels. Whether and how the factual information and its theoretical classification by science are preserved in the media, modified more or less strongly, or even completely distorted, varies from case to case. Appropriate action in environmental policy issues does not require a perfect information base. But it requires participants who strive for the best possible information and possess enough power of judgement to recognize obvious misinformation (fake news) as such, and who are aware of their responsibility.

17.3 The Problem of Power

Solutions to environmental and sustainability problems must be enforced, and anyone who wants to enforce something needs power. Power is often talked about, but what power is, is not easy to define. Power in the simplest sense is the ability of a person to achieve a self-set goal through an action. But political power always involves a multitude of people in a societal context. Accordingly, Max Weber writes "Power means every chance, within a social relationship, to enforce one's own will against resistance, no matter what this chance is based on". When speaking of power in the sphere of politics, one does not refer to an isolated action, but to an "ensemble of coordinated actions" (Foucault). Power is therefore the ability to influence such an ensemble, which is structurally always found in the form of institutions and organizations, and to steer it in a certain direction, in the case of transformations this is a direction that was not foreseen in the previous course of action. In our concepts, power in this sense belongs to the realm of the *Homo politicus* (see Chap. 5 on *Concepts of man*). The use of power never starts from scratch, but even the power for a new beginning presupposes existing power. This side of power would, in the language of our concepts, be addressed under the topic of *stocks* (see Chap. 14). To define power more precisely, one would have to distinguish this concept from the concept of violence or the concept of domination. The debate on how such distinctions should be made has been quite controversial in the political philosophy of the 20th century. Max Weber emphasizes the moment of always having to overcome resistance when exercising power, which is why he sees violence as an almost indispensable means for power. Hannah Arendt, on the other hand, emphasizes that the essence of power lies not in violence, but in the more or less willing participation of all those involved in the pending action ensemble.[1]

Achieving environmental and sustainability goals in the field of politics means gaining power for their enforcement and maintaining this power until these goals have been achieved. Such power, like all power, requires appropriate means: These include material resources, finances, networks, associations, administrations, organizations such as parties and NGOs, political alliances such as coalitions, public advocacy and advertising for the goals in all media, so that they are constantly present in societal discourse. Bringing together these means and aligning them towards a specific goal is one of the essential tasks of political actors. The organization, exercise and preservation of this power in the sense of *responsibility* (see Chap. 6) for these goals requires the *power of judgement* (see Chap. 7) of *Homines Politici*. To illustrate and justify what needs to be done in detail here far exceeds the possibilities of this book. However, from many years of experience in political consulting in the Federal Republic of Germany, Great Britain, the USA and the People's Republic of China, we would like to shed some light on this topic.

a) Those who advocate for far-reaching environmental protection and comprehensive sustainability goals are generally not on the side of the politically significant majority. Especially when a new issue arises, environmentalists are initially the weaker ones. This means that the power of the opponents is usually superior. Therefore, environmental movements are something like a "threatened species" at least in their beginnings. Since environmental protection, seen from the perspective of power, is usually on the weaker side, it depends on the institutions of law. Committed actors depend on the law protecting their freedom to express themselves publicly and to attract attention through demonstrations and other conspicuous measures, so that it is impossible to force them to give up their demands or even to give up entirely. They are often uncomfortable, even alien to some parts of society. Carefully balanced breaches of the law in individual cases can attract additional attention as symbolic actions. But this means must be used with discretion and a certain caution. A "five to twelve" mood could tempt one to accept even far-reaching breaches of the law, so that the rescue of the world from environmental and climate disasters might still succeed. This approach has so far proven to be ineffective. It is not uncommon for the breach of law to be discussed more strongly in public discourse than the reason why law was broken.

b) The use of power is always accompanied by a certain ambiguity. One is dependent on allies, with whose positions one does not agree in many respects, and one must make compromises that contain elements that one cannot actually approve of in principle. And one must develop a sense for where the limit is: Which allies are definitely out of the question, what kind of compromises are to be excluded at all costs? Often more problematic than the choice of questionable

[1] The considerations of this section refer to the article by Paulick (2018) in the Socialnet Lexicon.

allies, however, is fundamentalism in environmental matters, i.e., a mentality that cannot enter into any alliances at all, as it looks so much at the purity of the doctrine in every potential alliance that actually no one wants to join except those who are already involved. Anyone who wants to achieve sustainability policy goals cannot avoid dealing with power. This requires trust and hope: trust that the right and just cause will have its long-term effect on shaky allies and those with whom one has entered into uncomfortable compromises. And hope is needed to endure the ambiguities and misguided paths of political affairs even remotely, without despairing.

References

Paulick, C. (2018). *socialnet Lexikon: Macht | socialnet.de*. socialnet. https://www.socialnet.de/lexikon/Macht.

Piketty, T. (2022). *A Brief History of Equality*. Harward University Press.

MIX
Papier aus verantwortungsvollen Quellen
Paper from responsible sources
FSC® C105338

If you have any concerns about our products,
you can contact us on
ProductSafety@springernature.com

In case Publisher is established outside the EU,
the EU authorized representative is:
Springer Nature Customer Service Center GmbH
Europaplatz 3, 69115 Heidelberg, Germany

Printed by Libri Plureos GmbH
in Hamburg, Germany